EXAMKRACKERS 1001 Questions in MCAT Chemistry

 Published in the United States of America by Osote Publishing, New Jersey

ISBN 1-893858-22-7

2005 Edition

To purchase additional copies of this book, call 1-888-572-2536 or fax orders to 1-859-255-0066.

examkrackers.com

osote.com

audioosmosis.com

Acknowledgements

Scott wishes to thank all the students he has worked with over the years. They questioned every argument that was less than watertight, highlighted the topics and theories that are most confusing, and made it seem all worthwhile. Thanks also go to Pam Mills and Bill Sweeney in the Hunter College Chemistry Department, for knowing that students always come first and for realizing that chemistry shouldn't just be for chemists.

Jon wishes to thank his wife, Silvia, for her support, especially during the difficult times in the past and those that lie ahead.

Table of Contents

Significant Figures

1. How many significant figures are in 0.008010?

   A. 2
   B. 3
   C. 4
   D. 6

2. The following values of a sample of gas were measured in the lab: $P = 1.5$ atm; $V = 10.3$ L; and $T = 298.0$ K. Based on these measurements, a student calculated the number of moles using the equation $PV = nRT$. Which calculated value has the correct number of significant figures? (Note: the value of R is 0.08206 L atm K^{-1} mol^{-1})

   A. 0.63 mol
   B. 0.632 mol
   C. 0.6318 mol
   D. 0.63180 mol

Atoms

Refer to the hypothetical element E shown below to answer questions 3-18.

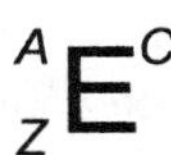

3. Z on element E represents:

   A. the number of neutrons.
   B. the number of protons.
   C. the number of neutrons plus protons.
   D. the number of electrons.

4. A on element E represents:

   A. the number of neutrons.
   B. the number of protons.
   C. the number of neutrons plus protons.
   D. the number of electrons.

5. C on element E represents:

   A. the number of electrons.
   B. the number of protons.
   C. the number of protons minus electrons.
   D. the number of neutrons plus protons.

6. A good approximation of the mass of one atom of element E in atomic mass units would be:

   A. A
   B. Z
   C. $A + Z$
   D. $A + Z + C$

7. If E represents any real element, which of the following statements must be true concerning the element E as shown above?

   A. The value of A is larger than Z.
   B. The value of Z is larger than A.
   C. If E is an ion, the value of Z is negative.
   D. The value of Z cannot be larger than the value of A.

8. If C for a sample of element E is zero, then:

   A. A and Z are equal.
   B. There are the same number of protons and neutrons in the sample.
   C. There are the same number of protons and electrons in the sample.
   D. The sample is ionized.

9. The atomic number on the element E is represented by:

   A. A
   B. Z
   C. C
   D. $A + Z$

10. The mass number on the element E is represented by:

    A. A
    B. Z
    C. C
    D. $A + Z$

11. Which of the following is always true of the relationship between A and Z on any stable element E?

    A. Z is greater than A.
    B. A is greater than Z.
    C. Z is exactly half as great as A.
    D. A minus Z gives the number of neutrons.

12. Which of the following could not be true for any given element E?

    A. There is more than one possible value for A.
    B. There is more than one possible value for C.
    C. There is more than one possible value for Z.
    D. There is more than one possible value for $A + Z$.

13. If two different atoms of element E have different values for A, they must be:

    A. different elements.
    B. ions of the same element.
    C. isotopes of the same element.
    D. isomers of the same element.

14. If *A* of element E has a value of 22, then:

A. *Z* must have a value of 11.
B. any form of element E must contain 22 electrons.
C. element E has an atomic weight of approximately 11 amu.
D. element E has an atomic weight of approximately 22 amu.

15. If *A* of element E has a value of 22, then one mole of element E:

A. has a mass of approximately 11 amu.
B. has a mass of approximately 22 amu.
C. weighs approximately 22 grams.
D. weighs approximately 22 x 6.02 x 10^{23} grams.

16. If *A* of element E has a value of 12, then a 24 gram sample of element E represents approximately:

A. ½ of an atom.
B. 2 atoms.
C. ½ mole of atoms.
D. 2 moles of atoms.

17. If one mole of a sample of element E weighs 14 grams, A of element E has a value of approximately:

A. 7
B. 14
C. 7 x 6.02 x 10^{23}
D. 14 x 6.02 x 10^{23}

18. The identity of element E can be attained with absolute certainty by knowing the value of which of the following?

A. Z
B. A
C. C
D. A + Z

19. Which of the following has a mass of approximately 1 amu?

A. one proton
B. one electron
C. one atom of ^{12}C
D. one mole of ^{12}C

20. Which of the following would have the greatest mass?

A. 25 protons, 25 neutrons, and 23 electrons
B. 24 protons, 25 neutrons, and 27 electrons
C. 25 protons, 24 neutrons, and 27 electrons
D. B and C would both have exactly the same mass, while A would be less

21. Two different isotopes of an element are isolated as neutral atoms. The atoms must have the same number of:

I. Protons
II. Neutrons
III. Electrons

A. I only
B. II only
C. I and III only
D. I, II, and III

22. Silicon exists as three different isotopes in nature. These are Si^{28}, Si^{29}, and Si^{30} with the atomic weights of 27.98 amu, 28.98 amu, and 29.97 amu respectively. Which isotope is likely to be most abundant in nature?

A. Si^{28}
B. Si^{29}
C. Si^{30}
D. They are nearly equal in abundance.

23. Arsenic (As) exists as a single isotope in nature. What is the expected number of neutrons in this isotope?

A. 33
B. 38
C. 42
D. 45

24. How much does a 3-mole sample of Na weigh?

A. 23 amu
B. 69 amu
C. 23 grams
D. 69 grams

25. How many atoms of Mg are in a 48 g sample of solid Mg?

A. 2
B. 4
C. 2 x 6.02 x 10^{23}
D. 4 x 6.02 x 10^{23}

26. The charge on one mole of electrons is given by Faraday's constant (*F* = 96,500 C/mol). What is the total charge of all the electrons in 2 grams of He?

A. 48,250 C
B. 96,500 C
C. 193,000 C
D. 386,000 C

27. Which of the following represents the charge on one mole of electrons?

A. 1 *e*
B. 6.02×10^{23} *e*
C. 1 C
D. 6.02×10^{23} C

28. How many neutrons are in one atom of the most common isotope of hydrogen?

A. 0
B. 1
C. 2
D. 4

29. Which of these elements, in its most common isotope, has more protons than neutrons in its nucleus?

A. H
B. He
C. C
D. U

30. Which element, in its most common isotope, has more neutrons than protons in its nucleus?

A. H
B. He
C. C
D. U

31. Which of the following is a complete list of the isotopes of carbon that are found in nature?

A. ^{14}C
B. ^{13}C, ^{14}C
C. ^{12}C, ^{13}C, ^{14}C
D. C, ^{12}C, ^{13}C, ^{14}C

32. Lithium occurs naturally in only two isotopic forms, lithium-6 and lithium-7. What can be said about the relative abundances of the two isotopes?

A. Lithium-6 is much more abundant
B. Lithium-7 is much more abundant
C. The isotopes are found in roughly equal abundances
D. The abundance oscillates as neutrons diffuse from one nucleus to another

33. 69% of naturally occurring copper is copper-63. If only one other isotope is present in natural copper, what is it?

A. copper-61
B. copper-62
C. copper-64
D. copper-65

34. The molecular formula for glucose is $C_6H_{12}O_6$. What is the mass of one mole of glucose?

72 12 96

A. 100 g
B. 180 g
C. 200 g
D. 360 g

The Periodic Table

35. Na and K belong to which family of elements?

A. the alkaline earth metals
B. the alkali metals
C. the transition metals
D. the representative elements.

36. The members of the noble gas family are sometimes called the:

A. royal elements
B. inert gases
C. representative elements
D. halogens

37. Alkaline earth metals generally form ions with a charge of:

A. +1
B. +2
C. –1
D. –2

38. When halogens make ions, they tend to:

A. lose one electron.
B. lose two electrons.
C. gain one electron.
D. gain two electrons.

39. Iron, silver, and mercury are:

A. representative elements.
B. transduction metals.
C. transition metals.
D. alkaline earth metals.

40. To which of the following families does magnesium belong?

A. the alkaline earth metals
B. the alkali metals
C. the transition metals
D. the representative elements.

41. Which of the following has naturally occurring ions with two different charges?

A. H
B. He
C. V
D. Sr

42. If X represents an alkali metal, and Y a halogen, what is the formula for the salt of X and Y?

A. XY
B. X_2Y
C. XY_2
D. The formula depends on X and Y

43. According to periodic trends, which of the following elements is expected to be the most malleable?

A. Au
B. Sn
C. C
D. Cu

44. The element with the greatest electronegativity is:

A. Cl
B. Fr
C. He
D. F

45. Which of the following elements is the most electronegative?

A. Be
B. Br
C. Cs
D. Kr

46. Energy of ionization is typically defined as:

A. the energy necessary to remove an electron from an element in its gaseous state.
B. the energy necessary to remove an electron from an element in its liquid state.
C. the energy necessary to add an electron to an element in its standard state.
D. the energy released when an element forms ions in aqueous solution.

47. Which of the following elements most easily accepts an extra electron?

A. Cl
B. Fr
C. He
D. Na

48. In a bond between any two of the following atoms, the bonding electrons would be most strongly attracted to:

A. Cl
B. Cs
C. He
D. I

49. Which of the following elements has the largest atomic radius?

A. Cl
B. Ar
C. K
D. Ca

50. Which of the following has the largest radius?

A. Cl^-
B. Ar
C. K^+
D. Ca^{2+}

51. Which of the following is ordered correctly in terms of atomic radius, from smallest to largest?

A. Al^{3+}, Al, S, S^{2-}
B. Al^{3+}, S, Al, S^{2-}
C. S, Al^{3+}, S^{2-}, Al
D. S, S^{2-}, Al^{3+}, Al

52. The attraction of the nucleus on the outermost electron in an atom tends to:

A. decrease moving from left to right and top to bottom on the periodic table.
B. decrease moving from right to left and top to bottom on the periodic table.
C. decrease moving from left to right and bottom to top on the periodic table.
D. decrease moving from right to left and bottom to top on the periodic table.

53. The greatest dipole moment is likely to be found in a bond where:

A. both bonding elements have high electronegativity.
B. both bonding elements have moderate electronegativity.
C. both bonding elements have low electronegativity.
D. one bonding element has high electronegativity and the other has low electronegativity.

54. Removing an electron from which of the following would require the most energy?

A. Na^-
B. Na
C. Na^+
D. Na^{2+}

55. An atom of phosphorous will be most similar in size to which of the following atoms?

A. O
B. Ge
C. As
D. Se

56. A naturally-occurring sample of which of the following has the smallest density at room temperature?

A. Aluminum
B. Magnesium
C. Sodium
D. Sulfur

57. A naturally-occurring sample of which of the following has the smallest density at room temperature?

A. Beryllium
B. Boron
C. Fluorine
D. Lithium

58. A naturally-occurring sample of which of the following has the smallest density at room temperature?

A. Carbon
B. Fluorine
C. Nitrogen
D. Oxygen

59. A naturally-occurring sample of which of the following has the smallest density at room temperature?

A. Argon
B. Chlorine
C. Phosphorus
D. Sulfur

60. Lithium's first and second ionization energies are 519 kJ/mol and 7300 kJ/mol, respectively. Element X has a first ionization energy of 590 kJ/mol and a second ionization energy of 1150 kJ/mol. Element X is most likely to be:

A. Oxygen
B. Sodium
C. Calcium
D. Xenon

61. Which of the following elements has chemical properties most similar to K?

A. Ca
B. Cs
C. Ar
D. O

62. Which of the following two elements are in the same family?

A. Cr and Fe
B. O and Se
C. B and C
D. Ce and Nd

63. Which compound listed below is likely to be the most similar chemically to table salt (NaCl)?

A. MgS
B. $NaNO_3$
C. AgCl
D. KBr

64. Removing an electron from which of the following would most likely require the most energy?

A. Na
B. Na^+
C. Mg
D. Mg^+

65. Removing an electron from which of the following would most likely require the most energy?

A. Cl
B. Cl^{2-}
C. H
D. Ca

66. The nucleus of which of the following would exert the greatest electrostatic force on its outermost electron?

A. Na
B. H
C. Cl
D. Mg

67. Many chemists consider the electronegativity of helium to be undefined. Why?

A. The small size of helium makes its electronegativity difficult to measure
B. Helium does not have inner-shell electrons
C. Helium atoms are electrically neutral
D. Helium does not form bonds with other elements

68. Which solution is most likely to be colored?

A. $Na_2CO_3(aq)$
B. NaCl(aq)
C. KBr(aq)
D. $FeCl_3(aq)$

69. Atom A and Atom B are in the same row of the periodic table. Atom A has a greater radius than atom B. Atom A probably also has a greater:

I. Electronegativity
II. First ionization energy
III. Atomic weight

A. III only
B. I and III
C. I, II, and III
D. None of the above

70. Because of the ease with which it is oxidized, pure sodium sometimes catches fire when exposed to water. Which of the following pure elements is most likely to catch fire when exposed to water?

A. Lithium
B. Beryllium
C. Magnesium
D. Potassium

71. Fe^{2+} has a higher ionization energy than Fe. Which of the following is a reasonable explanation for this fact?

A. Fe^{2+} is larger than Fe
B. Fe^{2+} is isoelectronic with chromium, which has a higher ionization energy than Fe
C. The outer electrons of Fe^{2+} experience a greater effective nuclear charge than those of Fe
D. Energy had to be put into Fe to ionize it to Fe^{2+}

72. Mg^{2+} is smaller than Na^{+}. Why?

A. Mg^{2+} has fewer electrons than Na^{+}, and the size of an ion is determined by the size of its electron cloud
B. Mg^{2+} has a greater mass than Na^{+}, and thus holds its electrons more tightly
C. Mg^{2+} has a greater atomic number than Na^{+}, and thus holds its electrons more tightly
D. Mg^{2+} has a smaller ionization energy than Na^{+}, and thus a smaller size

Molecules

73. How many electrons are in a single covalent bond?

A. 1
B. 2
C. 3
D. 4

74. As bond length between a given pair of atoms increases:

A. bond strength and bond energy decrease.
B. bond strength and bond energy increase.
C. bond strength increases and bond energy decreases.
D. bond strength decreases and bond energy increases.

75. When a bond is broken:

A. energy is always released.
B. energy is always absorbed.
C. energy is absorbed if the bond strength is positive.
D. energy is released if the bond strength is negative.

76. The force holding two atoms together in a chemical bond is:

A. the weak nuclear force.
B. the strong nuclear force.
C. electrostatic force.
D. gravitational force.

77. Two atoms are held together by a chemical bond because:

A. their nuclei attract each other.
B. the electrons forming the bond attract each other.
C. their nuclei are attracted to the bonding electrons.
D. the bonding electrons form an electrostatic cloud that wraps and contains both nuclei.

78. The distance between two nuclei in a chemical bond is determined by:

A. a balance between the repulsion of the nuclei for each other and the attraction of the nuclei for the bonding electrons.
B. the size of the protons.
C. the size of the neutrons.
D. the size of the electrons.

79. How does the energy of a typical carbon-carbon double bond compare to the energy of a typical carbon-carbon single bond?

A. The bond energy of the double bond is less than that of the single bond
B. The bond energy of the double bond is greater than that of the single bond, but less than twice as great
C. The bond energy of the double bond is twice that of the single bond
D. The bond energy of the double bond is more than twice as great as that of the single bond

80. In a water molecule, oxygen has a partial negative charge because:

A. oxygen is more electronegative than hydrogen
B. oxygen has more valence electrons than hydrogen
C. oxygen is sp^3 hybridized
D. water is bent

81. What is the mass percent of nitrogen in NO_2?

A. 14.0%
B. 20.5%
C. 30.4%
D. 33.3%

82. Which of the following has the longest bond between carbon and oxygen?

A. CO
B. CO_2
C. K_2CO_3
D. CH_3OH

83. How do the bond lengths in CS_3^{2-} compare?

A. two are the same length, the other is longer
B. two are the same length, the other is shorter
C. they are all different lengths
D. they are all the same length

84. In which of the following compounds does carbon have the greatest percent by mass?

A. CCl_4
B. CH_3OH
C. C_3H_7OH
D. $C_3H_7NH_2$

85. Which of the following compounds is not possible?

A. SF_6, because fluoride does not have empty d-orbitals available to form an expanded octet
B. OCl_6, because oxygen does not have empty d-orbitals available to form an expanded octet
C. H^-, because hydrogen forms only positive ions
D. PbO_2, because the charge on a lead ion is only +2

86. According to VSEPR theory, what is the molecular geometry of sulfur tetrafluoride?

A. Tetrahedral
B. square planar
C. seesaw
D. it depends on the relative electronegativity of sulfur and fluoride

87. Which of the following is true of all pure compounds?

A. They are each made from a single element.
B. They exist as a collection of separate and identical molecules.
C. The relative number of atoms of one element compared to another can always be represented by a ratio of two whole numbers.
D. They are held together by intermolecular bonds.

88. What is the mass of one molecule of water?

A. 18 g
B. 18 amu
C. 18 moles
D. 18 g/mol

89. What is the mass in kilograms of a single water molecule?

A. $(18)(6.02 \times 10^{23})(1000)$

B. $\dfrac{18}{(6.02 \times 10^{23})(1000)}$

C. $\dfrac{(18)(1000)}{6.02 \times 10^{23}}$

D. $\dfrac{18}{6.02 \times 10^{23}}$

90. What is the percent by mass of chlorine in carbon tetrachloride?

A. 50%
B. 75%
C. 80%
D. 92%

91. The mass percent of a compound is as follows: 71.65% Cl; 24.27% C; and 4.07% H. What is the empirical formula of the compound?

A. $ClCH_3$
B. $ClCH_2$
C. ClC_2H_5
D. Cl_2CH_2

92. The mass percent of a compound is as follows: 71.65% Cl; 24.27% C; and 4.07% H. If the molecular weight of the compound is 98.96, what is the molecular formula of the compound?

A. ClC_2H_2
B. $ClCH_2$
C. $Cl_2C_2H_4$
D. $Cl_3C_3H_8$

93. The mass percent of a compound is as follows: 43.64% P, and 56.36% O. What is the empirical formula of the compound?

A. PO
B. PO_2
C. P_2O_3
D. P_2O_5

94. The mass percent of a compound is as follows: 43.64% P, and 56.36% O. If the molecular weight of the compound is 283.88, what is the molecular formula of the compound?

A. P_2O_3
B. P_2O_5
C. P_3O_7
D. P_4O_{10}

95. When 7.0 grams of hydrated potassium iodide is heated, the result is 5.5 grams of anhydrous potassium iodide. What was the percent by mass of water in the hydrated potassium iodide?

A. 7.0%
B. 10%
C. 21%
D. 27%

96. Which of the following could be an empirical formula?

A. N_2O_4
B. H_2O
C. N_3O_6
D. $Na_2(PO_3)_2$

97. How many carbons are in 22 grams of CO_2?

A. 11 atoms
B. 0.5 mole
C. 1 mole
D. 2 moles

98. How many nitrogens are in 34 grams of NH_3?

A. 68 atoms
B. 0.5 mole
C. 1 mole
D. 2 moles

99. Consider the following reaction:

$$C_2H_5OH(l) + 3O_2(g) \rightarrow 2CO_2(g) + 3H_2O(g)$$

If 54 grams of water vapor are produced, how many moles of hydrogen atoms participated in the reaction?

A. 3
B. 6
C. 9
D. 18

100. Which of the following compounds is 25% nitrogen by mass?

A. NH_3
B. $NH(CH_3)(CHCH_2)$
C. $NH_2CH_2NHCH_2NHCH_2NHCH_3$
D. NO

101. Which of the following compounds is 25% nitrogen by mole fraction?

A. NH_3
B. $NH(CH_3)(CHCH_2)$
C. $NH_2CH_2NHCH_2NHCH_2NHCH_3$
D. NO

102. If the mass percent of nitrogen in a compound is 10% and there are two nitrogen atoms in each molecule of the compound, what is the molecular weight of the compound?

A. 28
B. 70
C. 140
D. 280

103. What is the empirical formula for benzene (C_6H_6)?

A. CH
B. C_2H_2
C. C_6H_6
D. C_6H_{12}

104. What is the empirical formula of acetic acid, CH_3COOH?

A. CH_4
B. CH_2O
C. $C_2H_4O_2$
D. CH_3COOH

105. The mass percent of a compound is as follows: 6% H; and 94% O. What is the empirical formula of the compound?

A. HO
B. H_2O
C. H_2O_2
D. H_3O_3

106. Which of the following is sufficient for determining the molecular formula of a compound?

A. The molecular weight of a compound
B. The percent by mass of a compound
C. The percent by mass and the empirical formula of a compound
D. The percent by mass and the molecular weight of a compound

107. The empirical formula of a hydrocarbon is known to be CH_2. What can be determined from this information?

A. The molar mass and percent composition of the hydrocarbon.
B. The molar mass of the hydrocarbon, but not its percent composition.
C. The percent composition of the hydrocarbon, but not its molar mass.
D. Neither the molar mass nor the percent composition of the hydrocarbon.

Chemical Reactions and Equations

108. A 2.0 kilogram block had dimensions 3 cm x 5 cm x 8 cm. What is its density?

A. 1.67×10^{-3} kg/m^3
B. 1.67 kg/m^3
C. 167 kg/m^3
D. 1.67×10^4 kg/m^3

109. The radius of the Earth's orbit about the sun is 1.5×10^8 km. Which expression could be used to find the average rate at which the Earth travels around the sun in meters per second?

A. $\dfrac{(60)(60)(24)}{(2)(3.14)(1.5\times10^8)(365)}$

B. $\dfrac{(2)(3.14)(1.5\times10^8)(60)(60)}{(1000)(365)(24)}$

C. $\dfrac{(2)(3.14)(1.5\times10^8)(60)}{(1000)(365)(24)(60)}$

D. $\dfrac{(2)(3.14)(1.5\times10^8)(1000)}{(365)(24)(60)(60)}$

110. Which of the following is a physical reaction?

A. boiling
B. combustion
C. dehydration
D. elimination

111. Which of the following bonds might be broken in a physical reaction?

A. hydrogen bonds
B. peptide bonds
C. covalent bonds
D. intramolecular bonds

112.

$$CH_4 + 2O_2 \rightarrow CO_2 + 2H_2O$$

The reaction shown above is NOT an example of a(n):

A. combustion reaction.
B. chemical reaction.
C. oxidation-reduction reaction.
D. physical reaction.

113. What is the balanced reaction for the combustion of methane?

A. $NH_3 + OH^- \rightarrow NH_4OH$
B. $CH_3OH + 2O_2 \rightarrow CO_2 + 2H_2O$
C. $CH_4 + OH^- \rightarrow CH_3OH$
D. $CH_4 + 2O_2 \rightarrow CO_2 + 2H_2O$

114. Which reaction gives the balanced reaction for the combustion of ethanol (C_2H_5OH)?

A. $C_2H_5OH + O_2 \rightarrow CO_2 + H_2O$
B. $C_2H_5OH + 2O_2 \rightarrow 2CO_2 + 2H_2O$
C. $C_2H_5OH + 3O_2 \rightarrow 2CO_2 + 3H_2O$
D. $4C_2H_5OH + 13O_2 \rightarrow 8CO_2 + 10H_2O$

115. Which of the following represents the balanced double displacement reaction between copper (II) chloride and iron (II) carbonate?

A. $Cu_2Cl + Fe_2CO_3 \rightarrow Cu_2CO_3 + Fe_2Cl$
B. $CuCl_2 + Fe(CO_3)_2 \rightarrow Cu(CO_3)_2 + FeCl_2$
C. $Cu_2Cl + FeCO_3 \rightarrow Cu_2CO_3 + FeCl$
D. $CuCl_2 + FeCO_3 \rightarrow CuCO_3 + FeCl_2$

116. The following is an unbalanced reaction:

$$C_{12}H_{22}O_{11}(l) + O_2(g) \rightarrow CO_2(g) + H_2O(g)$$

How many moles of oxygen gas are required to burn one mole of $C_{12}H_{22}O_{11}$?

A. 1
B. 6
C. 12
D. 24

117. The following is an unbalanced reaction:

$$C_6H_{12}O_6(s) + O_2(g) \rightarrow CO_2(g) + H_2O(g)$$

How many moles of oxygen gas are required to burn one mole of $C_6H_{12}O_6$?

A. 1
B. 3
C. 6
D. 12

118. The following is an unbalanced reaction for the combustion of hexane (C_6H_{14}):

$$C_6H_{14}(g) + O_2(g) \rightarrow CO_2(g) + H_2O(g)$$

How many moles of oxygen gas are required to burn 2 moles of hexane?

A. 6
B. 12
C. 14
D. 19

119. The following is an unbalanced reaction:

$$Fe(s) + O_2(g) \rightarrow Fe_2O_3(s)$$

If 2 moles of iron react to completion with 2 moles of oxygen gas, what remains after the reaction?

A. 1 mole of Fe_2O_3 only
B. 1 mole of Fe_2O_3 and 1/2 mole of oxygen gas
C. 1 mole of Fe_2O_3 and 1/2 mole of iron
D. 1 mole of Fe_2O_3, 1 mole of iron, and 1 mole of oxygen gas.

120. The following is an unbalanced reaction:

$$Au_2S_3(s) + H_2(g) \rightarrow Au(s) + H_2S(g)$$

If 1 mole of $Au_2S_3(s)$ is reacted with 5 moles of hydrogen gas, what is the limiting reagent?

A. $Au_2S_3(s)$
B. $H_2(g)$
C. $Au(s)$
D. $H_2S(g)$

121. Fifteen moles of $N_2O_4(l)$ are reacted with $N_2H_3(CH_3)(l)$ to produce 36 moles of water via the equation shown below:

$$5N_2O_4(l) + 4N_2H_3(CH_3)(l) \rightarrow 12H_2O(g) + 9N_2(g) + 4CO_2(g)$$

How many moles of $N_2H_3(CH_3)(l)$ are used up in the reaction?

A. 4
B. 8
C. 10
D. 12

122. Ten moles of $N_2O_4(l)$ are added to an unspecified amount $N_2H_3(CH_3)(l)$ according to the equation shown below:

$$5N_2O_4(l) + 4N_2H_3(CH_3)(l) \rightarrow 12H_2O(g) + N_2(g) + CO_2(g)$$

If 23 moles of water are produced and the reaction runs to completion, what is the limiting reagent?

A. $N_2O_4(l)$
B. $N_2H_3(CH_3)(l)$
C. $H_2O(g)$
D. There was no limiting reagent.

123. Given the following reaction:

$$N_2(g) + 3H_2(g) \rightarrow 2NH_3(g)$$

which of the following could NOT be true?

A. 15 moles of nitrogen gas react with 45 moles of hydrogen gas to form 30 moles of ammonia gas.
B. 5 molecules of nitrogen gas react with 15 molecules of hydrogen gas to form 10 molecules of ammonia gas.
C. 25 grams of nitrogen gas react with 75 grams of hydrogen gas to form 50 grams of ammonia gas.
D. 28 grams of nitrogen gas react with 6 grams of hydrogen gas to form 34 grams of ammonia gas.

124. In the following reaction, which is run at 600 K, 4.5 moles of nitrogen gas are mixed with 11 moles of hydrogen gas:

$$N_2(g) + 3H_2(g) \rightarrow 2NH_3(g)$$

The reaction produces 6 moles of ammonia. How many moles of nitrogen gas remain?

A. 0 mol
B. 1.5 mol
C. 3.0 mol
D. 4.5 mol

125. In the following reaction, which is run at 600 K, 4.5 moles of nitrogen gas are mixed with 11 moles of hydrogen gas:

$$N_2(g) + 3H_2(g) \rightarrow 2NH_3(g)$$

The reaction produces 6 moles of ammonia. What is the percent yield of ammonia?

A. 42%
B. 57%
C. 82%
D. 100%

126. When the following reaction is run, a 75% yield is achieved:

$$PCl_3(g) + 3NH_3(g) \rightarrow P(NH_2)_3(g) + 3HCl(g)$$

How many moles of phosphorous trichloride are required to produce 328 grams of HCl?

A. 1 mol
B. 2 mol
C. 3 mol
D. 4 mol

127. The following reaction is run to completion:

$$P_4(s) + 6Cl_2(g) \rightarrow 4\ PCl_3(l)$$

How much phosphorous is required to produce 275 g of phosphorous trichloride?

A. 62 g
B. 124 g
C. 248 g
D. 275 g

128. The following reaction is run to completion:

$$As_4O_6(s) + 6C(s) \rightarrow As_4(g) + 6CO(g)$$

How much carbon is required to produce 75 g of arsenic gas?

A. 12 g
B. 18 g
C. 48 g
D. 72 g

129. Given the following reaction:

$$2SO_2(g) + O_2(g) \rightarrow 2SO_3(g)$$

how many moles of oxygen are needed to produce 3.5 moles of sulfur trioxide?

A. 1.00 mol
B. 1.75 mol
C. 2.50 mol
D. 3.00 mol

Bonding in Solids

130. Which of the following most accurately describes ice?

A. crystalline solid
B. amorphous solid
C. polymer
D. ionic compound

131. Which of the following is false?

A. Crystalline solids have a sharp melting point.
B. Amorphous solids melt over a temperature range.
C. Rapid cooling of polymers is likely to result in an amorphous solid, while slow cooling is more likely to form crystals.
D. Any pure compound can be separated into specific repeating units called molecules.

132. Which of the following compounds lacks ionic bonds?

A. NaCl
B. NaH
C. HCl
D. $Ca_3(PO_4)2$

133. Which of the following substances forms a molecular crystalline solid?

A. diamond
B. table salt
C. gold
D. ice

134. Why are the intermolecular forces in water ice stronger than the intermolecular forces in dry ice (CO_2)?

A. Water can exist in the liquid phase at one atmosphere, while carbon dioxide cannot
B. Water has a much lower molecular weight than carbon dioxide
C. Water is polar, while carbon dioxide is nonpolar
D. Water molecules have less kinetic energy than carbon dioxide molecules

135. Diamond melts at a much higher temperature than water ice. Which of the following is an explanation for this fact?

A. Diamond is made entirely of carbon, which is a stronger atom than hydrogen or oxygen
B. Diamond is formed under higher pressure
C. Each atom in diamond is linked to four adjacent atoms by covalent bonds, while in water ice, the links between molecules are weaker hydrogen bonds
D. Diamond is nonpolar, while water is polar

136. Magnesium oxide melts at 2826°C; sodium fluoride melts at 996°C. Which of the following is a possible explanation for this fact?

A. Oxygen is more electronegative than fluorine, creating stronger bonds.
B. The charges on the ions in magnesium oxide are greater than the ions in sodium fluoride, leading to stronger ionic bonds.
C. Magnesium oxide exhibits relatively strong dipole-dipole bonding, while sodium fluoride exhibits only van der Waal's forces.
D. More energy is needed to randomize the crystal structure of magnesium oxide, because magnesium and oxygen ions are more massive than sodium and fluoride ions, respectively.

Quantum Numbers

137. How many quantum numbers are necessary to describe a single electron in an atom?

A. 1
B. 2
C. 3
D. 4

138. Which quantum number designates the shell level of an electron?

A. The principal quantum number
B. The azimuthal quantum number
C. The magnetic quantum number
D. The electron spin quantum number

139. Which of the following sets of quantum numbers describes the highest energy electron?

A. $n = 3; l = 2; m_l = 2; m_s = -1/2$
B. $n = 2; l = 1; m_l = 0; m_s = -1/2$
C. $n = 1; l = 0; m_l = 0; m_s = -1/2$
D. $n = 2; l = 1; m_l = 0; m_s = +1/2$

140. Only one set of the following quantum numbers could exist. Which set is it?

A. $n = 3; l = 3; m_l = 2; m_s = -1/2$
B. $n = 2; l = 1; m_l = 2; m_s = -1/2$
C. $n = 4; l = 2; m_l = 2; m_s = -1/2$
D. $n = 1; l = 2; m_l = 3; m_s = +1/2$

141. What is the maximum number of electrons that can fit in a shell with principal quantum number 3?

A. 2
B. 3
C. 10
D. 18

Heisenberg Uncertainty Principle

142. In what way is it inaccurate to picture an electron as a tiny particle orbiting a nucleus?

A. An electron is actually much larger than a typical nucleus
B. The electron jumps from orbit to orbit more frequently than would be predicted by classical mechanics
C. Since it is impossible to know both the position and momentum of an electron simultaneously, it is inappropriate to consider the electron to be a localized particle with a definite orbit
D. It is difficult to determine the precise orbit experimentally

143. Does the Heisenberg uncertainty principle apply to macroscopic objects such as basketballs?

A. Yes, but the large size of a basketball makes it difficult to determine its position with precision anyway
B. Yes, but the large mass of a basketball makes the uncertainty in velocity very small even if the uncertainty in position is also very small
C. No, because the basketball is made up of very many atoms, and the uncertainties cancel out
D. No, because a basketball is constantly interacting with its environment

Energy Level of Electrons

144. Which of the following is true of the energy levels for an electron in a hydrogen atom?

A. Since there is only one electron, that electron must be in the lowest energy level
B. The spacing between the n=1 and n=2 energy levels is the same as the spacing between the n=4 and n=5 energy levels
C. The energy of each level can be computed from a known formula
D. The energy levels are identical to the levels in the He^+ ion

145. Which of the following does not exist?

A. IBr
B. UF_6
C. OF_5
D. $NaLiCO_3$

146. Monatomic hydrogen gas is placed in a container of fixed volume, initially at STP. As the temperature is slowly raised, spectral lines corresponding to electrons in energy levels above the ground level appear. No matter how far the temperature is raised, however, no spectral lines for electrons above the $n = 7$ level are ever found. Which of the following is a possible explanation for this phenomenon?

A. No elements have electrons in levels above $n = 7$
B. Energy levels above $n = 7$ correspond to orbitals so large that the hydrogen atoms would overlap, disrupting the spectral lines
C. At the temperatures required to raise electrons to orbitals above $n = 7$, hydrogen nuclei would decompose
D. Beyond $n = 7$, STP cannot be maintained

147. An electron in a certain element can have energies of –2.3, –5.1, –5.3, –8.2, and –14.9 eV. –14.9 eV is the ground state of the electron, and no other levels exist between –14.9 and –2.3 eV. Which of the following represents a partial list of photon energies that could be absorbed by an electron in the ground state of this atom? All energies are in electron volts.

A. –2.3, –5.1, –5.3, –8.2, –14.9
B. 0.2, 2.8, 2.9, 6.7
C. 2.3, 5.1, 5.3, 8.2, 14.9
D. 6.7, 9.6, 9.8, 12.6, 15.0, 16.1

148. Suppose an element in its ground state is capable of absorbing photons with energy 2.3 eV and 4.1 eV, but no other intermediate energies. If the atom in its ground state absorbs a photon of energy 4.1 eV, it is found to sometimes later emit a single photon of 4.1 eV, but sometimes it emits a photon with energy:

A. 1.8 eV
B. 2.0 eV
C. 2.3 eV
D. 4.1 eV

149. If a sulfur atom is in its ground state, how many unpaired electrons does it have?

A. 0
B. 1
C. 2
D. 4

150. What is the electron configuration of a bromine atom in the ground state?

A. $1s^22s^22p^63s^23p^63d^{10}4s^24d^{10}4p^5$
B. $1s^22s^22p^63s^23p^64s^24d^{10}4p^5$
C. $1s^22s^22p^63s^23p^63d^{10}4s^24p^5$
D. $1s^22s^22p^63s^23p^64s^24p^5$

151. What is the electron configuration of iron in the ground state?

A. $[Ar]4s^24d^6$
B. $[Ar]3s^23d^6$
C. $[Ar]3s^24d^6$
D. $[Ar]3d^64s^2$

152. What is the electron configuration of the iodide ion in the ground state?

A. $[Kr]4d^{10}4d^{14}5s^25p^5$
B. $[Kr]3d^{14}4d^{10}5s^25p^6$
C. $[Kr]4d^{10}5s^25p^6$
D. $[Kr]5s^25p^5$

153. What is the electron configuration of the Cr^{3+} ion?

A. $[Ar]4s^23d^4$
B. $[Ar]4s^23d^1$
C. $[Ar]4s^23d^7$
D. $[Ar]3d^3$

154. Which of the following represents an excited state of an atom?

A. $1s^22s^23s^1$
B. $1s^22s^22p^1$
C. $1s^22s^22p^6$
D. $1s^22s^22p^63s^1$

155. Which of the following electron configurations could represent an excited state of an atom?

A. $1s^22s^22p^63s^23d^{10}$
B. $1s^22s^22p^63s^23p^64s^1$
C. $1s^22s^22p^63s^23d^53p^64s^2$
D. $1s^22s^22p^63s^23p^63d^{10}4s^24p^6$

156. Which of the following electron configurations could represent an ion?

I. $1s^22s^22p^6$
II. $1s^22s^22p^63s^2$
III. $1s^22s^22p^63s^23d^53p^6$

A. I only
B. I and II only
C. II and III only
D. I, II, and III

157. Suppose electrons could have *three* possible spin states (up, down, and sideways), rather than just two. Assuming nothing else was different, which of the following would be the correct ground state electron configuration for an element with atomic number 16?

A. $1s^21p^62s^22p^6$
B. $1s^22s^23s^23p^64s^24p^2$
C. $1s^32s^32p^93s^1$
D. $3s^23p^63d^8$

Gases

158. Which of the following is/are assumed in the kinetic-molecular theory of ideal gases?

I. The molecules of gas all move at the same speed
II. The molecules of gas all have negligible volume
III. The molecules of gas exert no attractive forces on each other

A. I only
B. I and III only
C. II and III only
D. I, II, and III

159. In the ideal gas law, what does the variable V represent?

A. The average speed of a gas molecule
B. The average velocity of a gas molecule
C. The volume of a gas molecule
D. The volume of the container which holds the gas

160. Which of the following affects the average force (per unit area) exerted by a gas on the wall of its container?

I. The average speed of a gas molecule
II. The frequency of collisions between gas molecules and the wall
III. The volume of a gas molecule

A. I only
B. II only
C. I and II only
D. I, II, and III

161. In an ideal gas, which of the following shows the relationship between pressure and temperature at constant volume?

A.

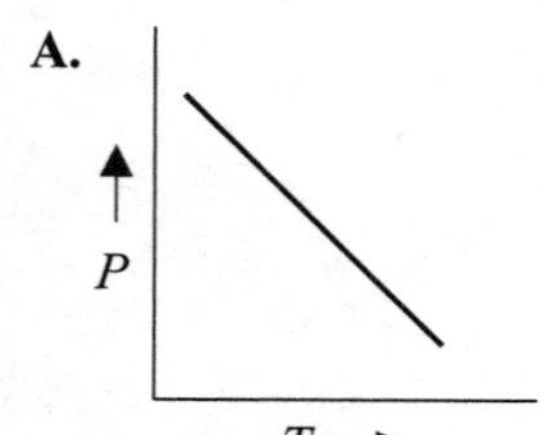

C.

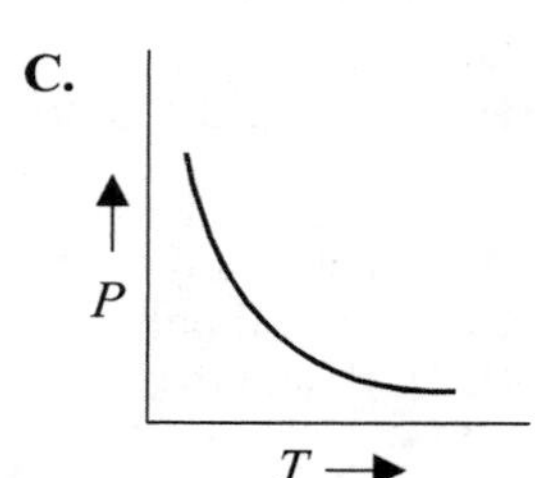

B.

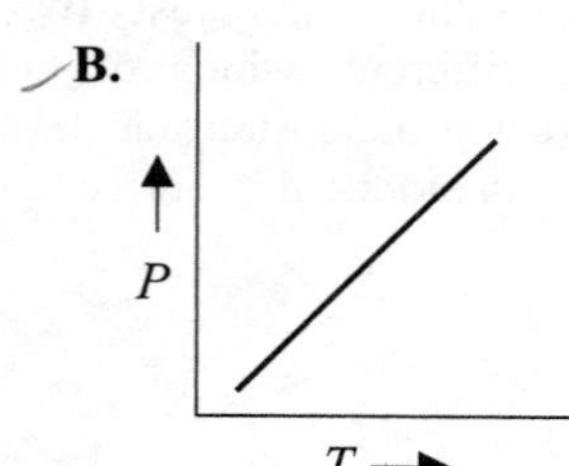

D.

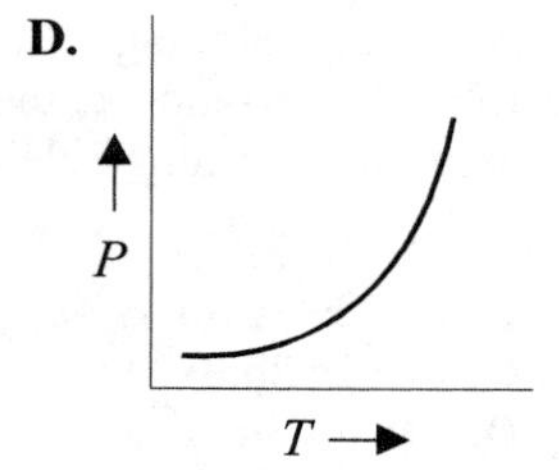

162. In an ideal gas, which of the following shows the relationship between pressure and volume at constant temperature?

A.

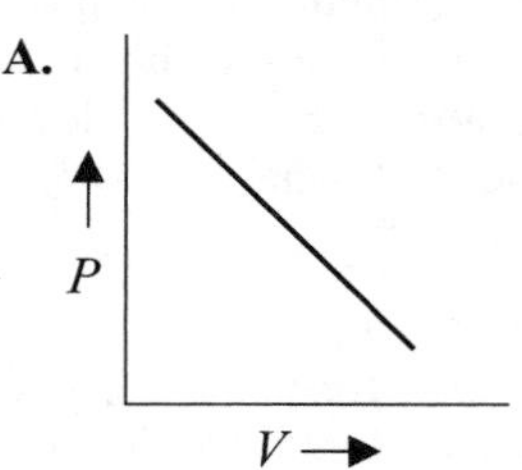

C.

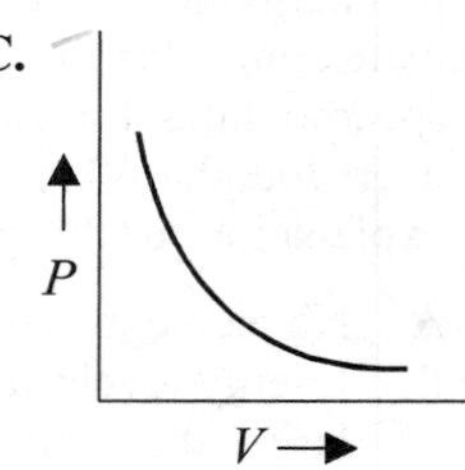

B.

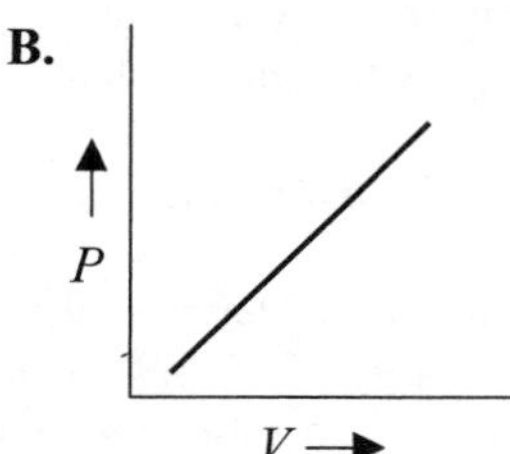

D.

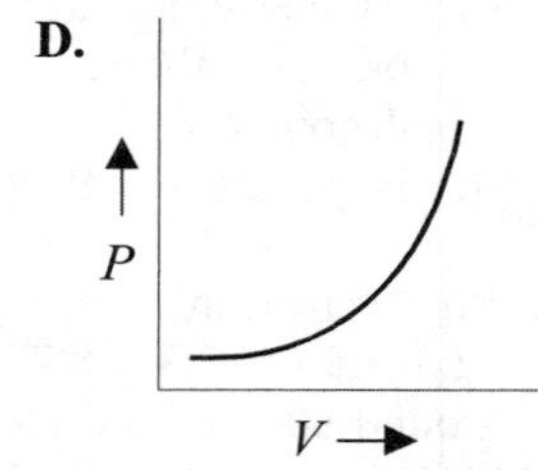

163. Container A contains gas at 300°C. Container B contains the same gas, but at 150°C. Which of the following is a true statement?

A. All of the gas molecules in container A move faster than all of the gas molecules in container B.
B. All of the gas molecules in container A move slower than all of the gas molecules in container B.
C. Each of the gas molecules in container A has more mass than each of the gas molecules in container B.
D. None of the above statements are true.

164. A gas initially fills a 2.0 L container. Heat is then added to the gas, raising its temperature from 300 K to 450 K, and increasing its pressure from 1.0 atm to 1.5 atm. What is the new volume of the gas?

A. 1.3 L
B. 2.0 L
C. 3.0 L
D. 4.5 L

165. A balloon has a volume of 500 mL when filled with gas at a pressure of 820 torr and a temperature of 300 K. How many moles of gas are inside the balloon? (R = 0.08206 L atm mol^{-1} K^{-1})

A. 0.022
B. 0.22
C. 2.2
D. 22

166. A balloon initially contains 20 grams of helium at a pressure of 1000 torr. After some helium is let out of the balloon, the new pressure is 900 torr, and the volume is half of what is was. If the temperature has not changed, how much helium is now in the balloon?

A. 9 grams
B. 10 grams
C. 11 grams
D. cannot be determined from the information given

167. Argon is the most common noble gas in the atmosphere at the surface of the Earth, despite the fact that helium is much more common in the universe. Why?

A. Argon is a byproduct of the decay of radioactive xenon in the Earth's crust
B. Argon is the most stable of the noble gasses
C. The production of argon is catalyzed by greenhouse gases
D. Argon has a similar density to nitrogen, the major component of the atmosphere

168. A researcher wishes to have two 20.0 L chambers containing equal amounts (moles) of oxygen gas, one at 5 atm and the other at 3 atm. Is this possible?

A. Yes, if the oxygen samples in the containers have different densities
B. Yes, if the oxygen samples in the containers are at different temperatures
C. No, because once the volume and the number of moles of an ideal gas are known, all other parameters are determined
D. No, because volume and pressure are inversely proportional for an ideal gas

169. A researcher wishes to have two samples of oxygen gas with the same pressure, volume, and temperature, but with different densities. Is this possible?

A. Yes, because density is not given in the ideal gas law.
B. Yes, if different isotopes of oxygen are used.
C. No, because once pressure, volume, and temperature of an ideal gas are known, all other parameters are determined
D. No, because density is inversely proportional to volume

170. A researcher wishes to use two identical containers, one to store 5 moles of oxygen, and one to store 3 moles of nitrogen. Both are kept at 300 K and 5 atm. Is this possible?

A. Yes, since oxygen and nitrogen have different molar masses
B. Yes, since these conditions are not STP
C. No, because once pressure, volume, and temperature of an ideal gas are known, the number of moles is determined
D. No, because number of moles and temperature are directly proportional

171. An ideal gas is placed in a 3.0 L container with a piston. The pressure of the gas is initially 850 torr. How much *additional* pressure must be exerted on the piston in order to lower the volume of the container to 1.0 L? Assume the temperature of the gas does not change.

A. 283 torr
B. 850 torr
C. 1700 torr
D. 2550 torr

172. Gas A is at 25°C and 1 atmosphere. If the pressure is increased to 3 atmospheres without changing the volume, the new temperature will most likely be:

A. -174 °C
B. 8.3 °C
C. 28 °C
D. 621°C

173. An ideal gas with pressure 2 atmospheres is expanded to twice its initial volume. What is the new pressure?

A. 1 atm
B. 2 atm
C. 4 atm
D. Cannot be determined from the information given

174. Which of the following could be used to estimate the value of absolute zero?

A. Plot pressure vs. temperature values for an ideal gas at constant volume, and extrapolate the resulting line to low temperatures. The intercept with the temperature axis is an estimate of absolute zero.
B. Plot volume vs. temperature values for an ideal gas at constant pressure, and extrapolate the resulting line to low temperatures. The intercept with the temperature axis is an estimate of absolute zero.
C. Allow both volume and pressure to vary, and plot the product of pressure and volume vs. temperature. The intercept with the temperature axis is an estimate of absolute zero.
D. All of the above techniques would work

175. 50.0 grams of oxygen are placed in an empty 10.0 liter container at 28°C. Compared to an equal mass of hydrogen placed in an identical container (also at 28°C), the pressure of the oxygen is:

A. Less than the pressure of the hydrogen
B. Equal to the pressure of the hydrogen
C. Greater than the pressure of the hydrogen
D. Cannot be determined from the information given

176. Container A and B both contain 1.00 L of a gas at STP, but container A contains oxygen, while container B contains nitrogen. Assuming the gases behave ideally, the gases has the same:

I. number of molecules
II. density
III. kinetic energy

A. II only
B. III only
C. I and III only
D. I, II, and III

177. A container appears to be marked "Ne, 3.5 moles," but has no pressure gauge. By measuring the temperature and checking the volume of the container, the scientist uses the ideal gas law to estimate the pressure inside the container. Unfortunately, the handwriting on the canister is difficult to read, and the container actually contains 3.5 moles of He, not Ne. How does this affect the scientist's estimate of the pressure inside the container?

A. The estimate is too low
B. The estimate is too high
C. The estimate is correct, but only because both gases are monatomic
D. The estimate is correct, because the identity of the gas is irrelevant

178. A container contains only oxygen, nitrogen, carbon dioxide, and water vapor. If, at STP, the partial pressure of oxygen is 200 torr, carbon dioxide 10 torr, and water vapor 8 torr, what is the partial pressure of the nitrogen?

A. 14 torr
B. 542 torr
C. 760 torr
D. 782 torr

179. Consider the synthesis of ammonia from its constituent elements:

$$N_2 + 3H_2 \rightarrow 2NH_3$$

Which of the following expressions can be used to find the number of grams of ammonia that can be produced from 24.5 mL of hydrogen at STP, if the percent yield is 35%?

A. $\dfrac{(0.001)(24.5)(2/3)(17.0)(0.35)}{22.4}$

B. $\dfrac{(0.001)(24.5)(2/3)(17.0)}{(22.4)(0.35)}$

C. $\dfrac{(1000)(24.5)(2/3)(17.0)(0.35)}{22.4}$

D. $\dfrac{(1000)(24.5)(2/3)(17.0)}{(22.4)(0.35)}$

180. A sealed container initially contains solid carbon and pure oxygen gas. In the presence of a spark, all of the carbon is converted into a gas with density 1.25 g/L at STP. What is the identity of the gas?

A. C_2
B. CO
C. CO_2
D. Cannot be determined from the information given

181. When methane gas and oxygen gas are made to undergo combustion in a sealed container, and the temperature brought back to the original temperature, it is found that the pressure has not changed. Which of the following is consistent with the data?

A. Methane was initially in excess; the products of the reaction were carbon monoxide and water vapor, and some of the methane did not react
B. Methane and oxygen were present in amounts such that both were entirely consumed, producing carbon monoxide and water vapor
C. Oxygen was initially in excess; the products of the reaction were carbon dioxide and water vapor
D. Oxygen was initially in excess, the products of the reaction were carbon dioxide, hydrogen gas, and ozone

182. If 64.0 grams of oxygen gas at 5.0 atm pressure occupies 3.0 L, what is the temperature of the gas? (R = 0.08206 L atm/mol K)

A. 3 K
B. 91 K
C. 298 K
D. Cannot be determined from the information given.

183. The mole fraction of nitrogen in air is approximately 0.8. At STP, what is the partial pressure of nitrogen in air?

A. 608 torr
B. 760 torr
C. 800 torr
D. Cannot be determined from the information given

184. If the partial pressure of hydrogen in a container held at 5 atmospheres pressure is 35 torr, what is the mole fraction of hydrogen in the container?

A. 0.009
B. 0.046
C. 0.23
D. Cannot be determined from the information given

185. If the partial pressure of carbon dioxide is 30 torr at STP, what is the mass of the carbon dioxide present?

A. 0.04 grams
B. 1.7 grams
C. 44 grams
D. Cannot be determined from the information given

186. If the partial pressure of carbon dioxide is 30 torr at STP, what is the percent by mass of carbon dioxide?

A. 1.7%
B. 4%
C. 44%
D. Cannot be determined from the information given

187. If the partial pressure of carbon dioxide is 30 torr at STP, and nitrogen is the only other gas present, what is the percent by mass of carbon dioxide?

A. 4%
B. 6%
C. 61%
D. Cannot be determined from the information given

188. If the partial pressure of carbon dioxide is 30 torr at STP, the gas is 10% carbon dioxide by mass, and there is only one other species of gas present, which of the following could be the other species of gas?

A. Hydrogen
B. Methane
C. Oxygen
D. Chlorine

189. In an 11.2 liter container the partial pressure of nitrogen gas is 0.5 atmospheres at 25°C. What is the mass of nitrogen in the container?

A. 3.5 g
B. 7 g
C. 14 g
D. 28 g

190. A 5 Newton weight is placed on a piston with an 8-centimeter radius. The piston compresses helium in a sealed container until the piston stops moving because of the increase in pressure inside the container. The pressure in the container is now recorded. Some helium is then removed from the container, causing the piston to fall so that the volume of the container drops by 25%, at which point the gas can again support the piston, and the pressure is again recorded. How do the two recorded pressures compare?

A. Since pressure and volume are directly proportional, the second pressure is lower.
B. Since pressure and volume are inversely proportional, the second pressure is higher.
C. Since the force on the piston is constant, the pressures are the same.
D. There is not enough information to answer this question.

191. Which of the following formulas gives the kinetic energy of n moles of gas?

A. $1/2\ \mathcal{M}V^2$, where $\mathcal{M}$ is the molar mass of the gas and V is the volume of the container
B. $3/2\ nRT$, where n is the number of moles of gas, R is the ideal gas constant, and T is the absolute temperature
C. nPA, where n is the number of moles of gas, P is the total pressure, and A is the surface area of the container walls
D. $1/2\ nPA$, where n is the number of moles of gas, P is the total pressure, and A is the surface area of the container walls

192. Which of the following would be a good method for distinguishing between ordinary hydrogen and deuterium (a rare isotope of hydrogen)?

I. Measure the density of the gas at STP
II. Measure the rate at which the gas effuses
III. Determine the number of grams of gas that will react with one mole of oxygen to form water

A. None of the above
B. II only
C. III only
D. I, II, and III

193. What is the ratio of the average speed of an atom of neon to a molecule of hydrogen at the same temperature and pressure?

A. 1:10
B. 1:3.2
C. 3.2:1
D. 10:1

194. What is the ratio of the average speed of an atom of neon to another atom of neon at twice the temperature but the same pressure?

A. 1:1
B. 1:1.4
C. 1:2
D. 1:4

195. What is the ratio of the average speed of an atom of neon to another atom of neon at the same temperature but twice the pressure?

A. 1:1
B. 1:1.4
C. 1:2
D. 1:4

196. If an equimolar mixture of oxygen, nitrogen, krypton, and carbon dioxide are placed in a 10 L container with a one-micron (0.001 mm) wide hole in the side, so that the gases slowly leak out, which gas will predominate after 20% of the original gases have leaked out?

A. Oxygen
B. Nitrogen
C. Krypton
D. Carbon dioxide

197. A gas with empirical formula CH is found to take about 40% more time to diffuse the same distance as neon. What is the molecular formula of the gas?

A. CH_4
B. C_2H_2
C. C_3H_3
D. C_4H_4

198. Which of the following is equivalent to the ideal gas law constant, R? (k_b is Boltzmann's constant and N_A is Avogadro's number)

A. k_bN_a
B. k_b/N_a
C. N_a/k_b
D. K_b+N_a

199. The average molecular kinetic energy of a sample of carbon dioxide is found to be the same as that of a sample of neon. How do the temperatures of the samples compare?

A. The carbon dioxide is at roughly four times the temperature of the neon.
B. The neon is at roughly four time the temperature of the carbon dioxide.
C. The temperatures of the two gases are the same.
D. The temperatures cannot be determined from the information given.

200. The average speed of the molecules in a sample of carbon dioxide is found to be the same as that of a sample of neon. How do the temperatures of the samples compare?

A. The temperature of the carbon dioxide is roughly two times the temperature of the neon
B. The temperature of the neon is roughly two times the temperature of the carbon dioxide
C. The temperatures of the two gases are the same
D. Cannot be determined from the information given

Use the following information to answer questions 201-204. A cotton ball moistened with ammonia and a cotton ball moistened with aqueous HCl solution are inserted into either end of a glass tube 10 cm long. HCl and NH_3 gas move through the test tube. Where the gases meet, a circle of ammonium chloride precipitates inside the tube.

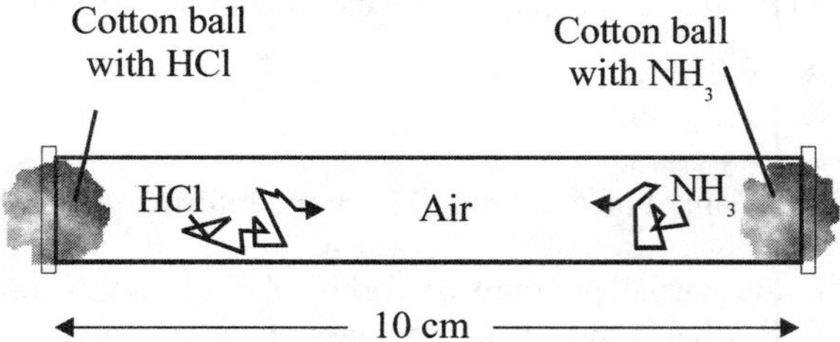

201. What is the approximate ratio of the average speed of the HCl gas molecules compared to the average speed of the NH_3 gas molecules?

A. 1:1
B. 1:1.5
C. 1.5:1
D. 36:17

202. The movement of gas in this experiment is called:

A. diffusion
B. effusion
C. extrusion
D. natural transport

203. The HCl molecules in this experiment don't follow a straight line, but follow a path more closely resembling that shown in the diagram. Why?

A. The natural spinning and rotation of HCl molecules causes them to change directions.
B. The HCl molecules bump into air molecules.
C. The HCl molecules bump into NH_3 molecules.
D. Any gas molecule, if acted upon by no outside forces, will follow a mean free path resembling the one shown in the diagram. This is called Brownian motion.

204. How far from the left end of the glass tube will the precipitate form?

A. 2 cm
B. 4 cm
C. 5 cm
D. 6 cm

Questions 205-208 refer to the apparatus shown below. A small pinhole is located between Stopcocks A and B.

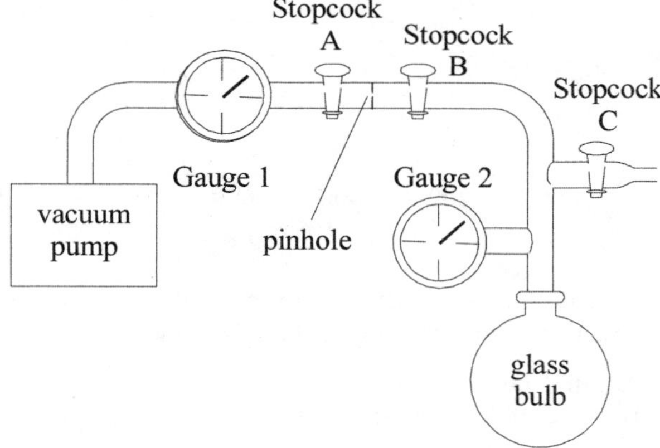

205. Stopcock B is closed. A gas sample is introduced into the glass bulb through an open Stopcock C. Stopcock C was then closed and the vacuum pump is turned on. In order to find the rate of effusion for a gas sample in the glass bulb, which of the following is true?

A. Readings should be taken from gauge 1 at regular intervals with stopcocks A and B closed.
B. Readings should be taken from gauge 1 at regular intervals with stopcocks A and B open.
C. Readings should be taken from gauge 2 with stopcocks A and B closed.
D. Readings should be taken from gauge 2 with stopcocks A and B open.

206. Which of the following might represent a series of readings taken from gauge 2 during the effusion of a sample from the glass bulb?

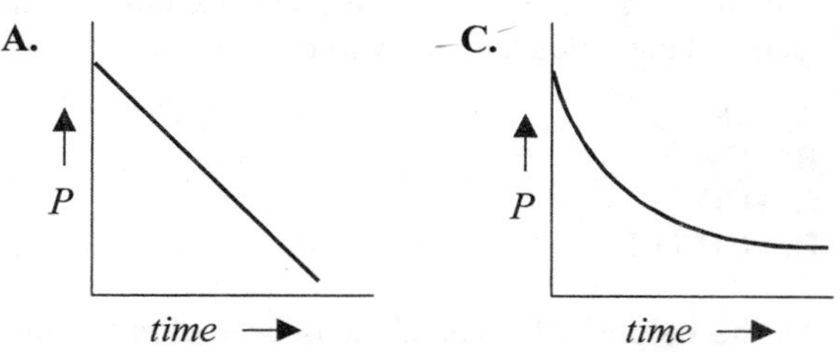

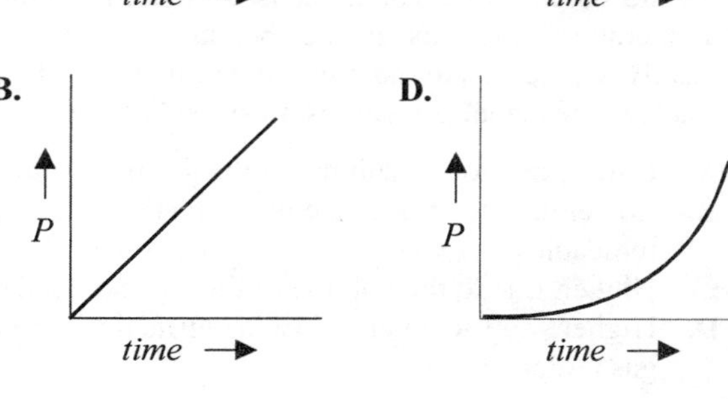

207. Under ideal conditions, which of the following would affect the effusion rate of the sample gas in the glass bulb?

I. The molecular weight of the sample gas
II. The pressure difference as measured by Gauge 1 and 2
III. The size of the gas molecules

A. I only
B. II only
C. I and II only
D. I, II, and III

208. The apparatus was used to test the rates of effusion of H_2, He, N_2 gases. Which of the following graphs most accurately reflects the results as measured by Gauge 2?

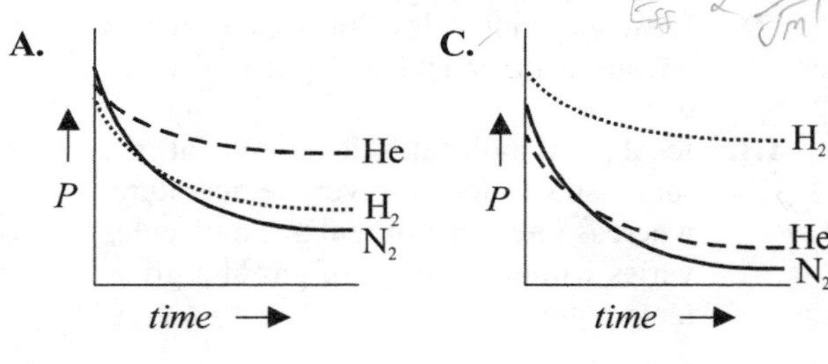

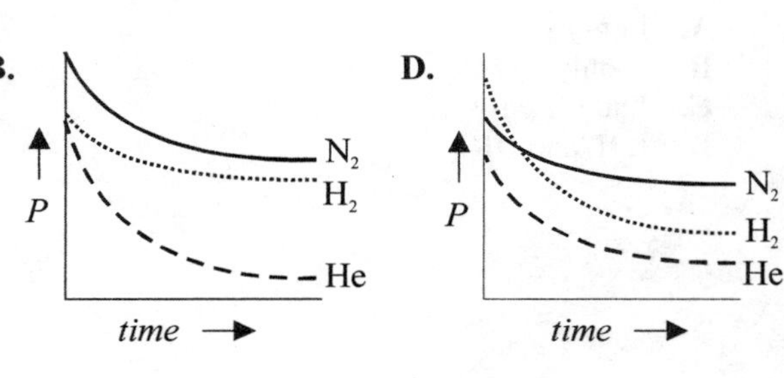

Real Gases

209. Which of the following gases, when compared under identical conditions, is likely to be least ideal?

A. O_2
B. O_3
C. CO_2
D. CH_3OH

210. As the volume of a container is decreased at constant temperature, the gas inside begins to behave less ideally. Compared to the pressure predicted by the ideal gas law, the actual pressure is most likely to be:

A. Lower, due to the volume of the gas molecules
B. Lower, due to intermolecular attractions among gas molecules
C. Higher, due to the volume of the gas molecules
D. Higher, due to intermolecular attractions between gas molecules

211. As the temperature of a sample of gas is decreased at constant volume, the gas inside begins to behave less ideally. Compared to the pressure predicted by the ideal gas law, the actual pressure is most likely to be:

A. Lower, due to the volume of the gas molecules
B. Lower, due to intermolecular attractions among gas molecules
C. Higher, due to the volume of the gas molecules
D. Higher, due to intermolecular attractions between gas molecules

212. Which of the following explains the pressure and volume deviations in a real gas compared to an ideal gas?

I. Ideal gas molecules don't have volume and real gas molecules do.
II. Ideal gas molecules don't exert forces on one another and real gas molecules do.
III. Ideal gas molecules all move at the same speed for a given temperature, whereas the speed of real gas molecules varies within a sample of gas at a given temperature.

A. I only
B. II only
C. I and II only
D. I, II, and III

Refer to the graph below to answer questions 213-219. The graph below shows PV/RT versus pressure for 1 mole of several gases at 300 K.

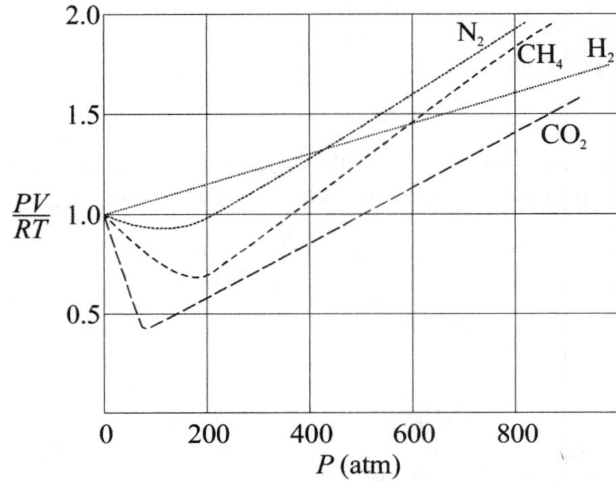

213. At 100 atm, CO_2 deviates from ideal behavior. The direction of this deviation is best explained by:

A. molecular volume
B. intermolecular attractions
C. temperature
D. molecular shape

214. At extremely high pressures, all gases deviate from the ideal gas law in the same direction. The greatest contributing factor to this deviation is:

A. molecular volume
B. intermolecular attractions
C. temperature
D. molecular shape

215. Deviations from ideal behavior typically increase with which of the following molecular characteristics?

A. molecular mass only
B. molecular complexity only
C. molecular mass and complexity
D. Deviations from ideal behavior depend upon temperature and pressure and are independent of molecular characteristics.

216. According to the graph, at what pressures of CO_2 is the ideal gas law correct and the proportionality constant R equal to 0.08206 L atm K^{-1} mol^{-1}?

A. 0 atm and 500 atm
B. 75 atm
C. 500 atm
D. 830 atm

217. According to the graph, which of the following would be true, if the ideal gas law, $PV = nRT$, were used to calculate the volume of a sample of CH_4 gas from measured variables at 200 atm and 300 K and again at 600 atm and 300 K?

A. The calculated volume would be less than the real volume for both calculations.
B. The calculated volume would be greater than the real volume for both calculations.
C. The calculated volume would be less than the real volume for the 200 atm sample and greater than the real volume for the 600 atm sample.
D. The calculated volume would be greater than the real volume for the 200 atm sample and less than the real volume for the 600 atm sample.

218. The data graphed for CO_2 actually pertain to a temperature of 313 K. The best explanation for this is:

A. CO_2 liquefies under low pressure at 300 K.
B. CO_2 liquefies under high pressure at 300 K.
C. At 300 K CO_2 behaves like an ideal gas.
D. At 300 K CO_2 deviates from ideal behavior.

219. Which of the following is indicated by the graph?

A. If temperature is sufficiently low, deviations due to molecular attractions dominate.
B. If temperature is sufficiently low, deviations due to molecular volume dominate.
C. If pressure is sufficiently high, deviations due to molecular attractions dominate.
D. If pressure is sufficiently high, deviations due to molecular volume dominate.

Refer to the graph below to answer questions 220-223. The graph below shows PV/RT versus pressure for 1 mole of nitrogen gas at three different temperatures T_1, T_2, and T_3.

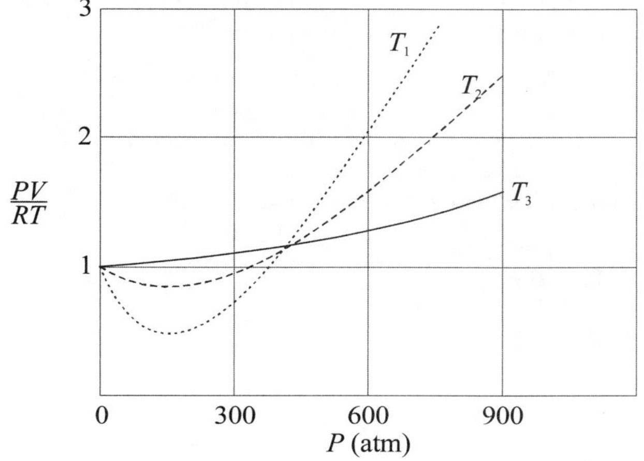

220. Which of the following gives the temperatures in increasing order?

A. $T_1 < T_2 < T_3$
B. $T_3 < T_2 < T_1$
C. $T_2 < T_1 < T_3$
D. $T_1 < T_3 < T_2$

221. At which temperature does the behavior of nitrogen most resemble that of an ideal gas?

A. T_1
B. T_2
C. T_3
D. The behavior of nitrogen resembles an ideal gas at all three temperatures.

222. If 2 moles of nitrogen had been used instead of 1 mole, what would be the approximate value of PV/RT at T_1 and 600 atm?

A. 1
B. 2
C. 3
D. 4

223. Based on the graph, which of the following properties of the nitrogen molecules plays the greatest role in explaining the deviation from ideal behavior for nitrogen gas at high temperatures?

A. polarity
B. molecular volume
C. intermolecular attractions
D. The answer cannot be deduced from the graph.

Use the van der Waals equation given below to answer questions 224-227. The van der Waals equation is used to predict the behavior of real gases. a and b are constants specific for a particular gas. Their values can be obtained by experiment or from a reference book.

$$\left(P + \frac{n^2a}{V^2}\right)(V - nb) = nRT$$

van der Waals Equation

224. The van der Waals constants a and b tend to increase when which of the following changes are made to the gas molecules?

A. Increasing both molecular mass and structural complexity
B. Decreasing both molecular mass and structural complexity
C. Increasing molecular mass but decreasing structural complexity
D. Decreasing molecular mass but increasing structural complexity

225. The constant *a* and *b* have positive values. What is the expected difference between using the ideal gas law, $PV = nRT$, to calculate the pressure from measured variables, and using van der Waals equation to calculate pressure from measured variables?

A. The pressure will always be lower when calculated with van der Waals equation.
B. The pressure will always be greater when calculated with van der Waals equation.
C. The pressure will always be the same when calculated with van der Waals equation.
D. Whether the value for pressure is lower or greater when calculated with van der Waals equation will depend upon the real volume and temperature.

226. Which of the following would be expected to have the greatest value for *a* and *b*?

A. He
B. Ne
C. Ar
D. Kr

227. Which of the following is true concerning the constants *a* and *b*?

A. A large value for *a* indicates *strong* intermolecular attractions, and a large value for *b* indicates a *large* molecular volume.
B. A large value for *a* indicates *weak* intermolecular attractions, and a large value for *b* indicates a *small* molecular volume.
C. A large value for *b* indicates *strong* intermolecular attractions, and a large value for *a* indicates a *large* molecular volume.
D. A large value for *b* indicates *weak* intermolecular attractions, and a large value for *a* indicates a *small* molecular volume.

Kinetics

228. Consider the complete combustion of ethanol (C_2H_6O) to form carbon dioxide and water. If the ethanol is consumed at a rate of 2.0 M s^{-1}, what is the rate at which carbon dioxide is produced?

A. 1.0 M s^{-1}
B. 2.0 M s^{-1}
C. 4.0 M s^{-1}
D. Cannot be determined from the information given

229. If the rate law for a reaction is:

$$\text{Rate} = k[A]^2[B]^4$$

what is the order of the reaction?

A. 2
B. 3
C. 4
D. 6

230. If the rate law for a reaction is:

$$\text{Rate} = k[A][B]^2$$

what is the order of the reaction?

A. 2
B. 3
C. 4
D. 6

231. If the rate law for a reaction is:

$$\text{Rate} = k[A]^2[B]^4$$

what is the order of A in this reaction?

A. 2
B. 3
C. 4
D. 6

232. The rate law for a reaction is:

$$\text{Rate} = k[A]^2$$

If B is also a reactant, what is the order of B in this reaction?

A. 0
B. 1
C. 2
D. Cannot be determined from the information given

233. Which of the following methods could be used to determine the rate law for a reaction?

I. Measure the initial rate of the reaction for a variety of reactant concentrations
II. Graph the concentration of the reactants as a function of time
III. Find the mechanism of the reaction

A. I only
B. III only
C. II and III only
D. I, II, and III

234. Consider the dissociation of hydrogen gas:

$$H_2(g) \rightarrow 2H(g)$$

Rates were measured for a number of different concentrations:

$[H_2]$	Rate/M s^{-1}
1.0	1.2 x 10^4
1.5	2.7 x 10^4
2.0	4.8 x 10^4

What is the rate law for this reaction?

A. Rate = $k[H_2]$
B. Rate = $k[H_2]^2$
C. Rate = $k[H]^2/[H_2]$
D. The rate law cannot be determined from the information given

235. Consider the reaction

$$2H_2 + 2NO \rightarrow N_2 + 2H_2O$$

Rates were measured for a number of different concentrations:

$[H_2]$	[NO]	Rate/M s^{-1}
0.1	0.3	230
0.2	0.3	460
0.3	0.1	80
0.2	0.1	50

What rate law is most consistent with this data?

A. Rate=$k[H_2][NO]$
B. Rate=$k[H_2][NO]^2$
C. Rate=$k[H_2]^2[NO]$
D. Rate=$k[H_2]^2[NO]^3$

236. Suppose a certain reaction has the rate law:

$$\text{Rate} = k[A]^{1/2}[B]$$

Which of the following can be concluded about the reaction?

A. Two molecules of B react for every molecule of A
B. Two molecules of A react for every molecule of B
C. B reacts at twice the rate of A
D. This reaction does not take place in a single step

Questions 237–242 depend on the following mechanism for the conversion of 2-iodo-2-methylpropane into an alcohol:

2-iodo-2-methylpropane $\rightarrow$
2-methylpropane cation + iodide

2-methylpropane cation + water $\rightarrow$
protonated 2-methylpropanol

protonated 2-methylpropanol + water $\rightarrow$
2-methylpropanol + hydronium

237. In the mechanism shown, iodide is a(n):

A. Reactant
B. Product
C. Intermediate
D. Catalyst

238. In the mechanism shown, 2-methylpropane cation is a(n):

A. Reactant
B. Product
C. Intermediate
D. Catalyst

239. In the mechanism shown, water is a(n):

A. Reactant
B. Product
C. Intermediate
D. Catalyst

240. In the mechanism shown, hydronium is a(n):

A. Reactant
B. Product
C. Intermediate
D. Catalyst

241. Which of the following is the correct net equation for the reaction shown above?

A. 2-iodo-2-methylpropane $\rightarrow$ 2-methylpropanol

B. 2-iodo-2-methylpropane $\rightarrow$
2-methylpropanol + hydronium

C. 2-iodo-2-methylpropane $\rightarrow$
2-methylpropanol + iodide

D. 2-iodo-2-methylpropane + $2H_2O \rightarrow$
2-methylpropanol + iodide + hydronium

242. If the reaction is first order in 2-iodo-2-methylpropane, this tells us that:

A. The first step is most likely the slowest
B. The first step is most likely the fastest
C. The reaction would give a higher yield in the presence of a catalyst
D. The reaction is not reversible

Questions 243-246 depend on the following mechanism for ozone destruction by NO gas:

$$NO + O_3 \rightarrow NO_2 + O_2$$
$$NO_2 + O \rightarrow NO + O_2$$

243. In the mechanism shown, NO is a(n):

A. Reactant
B. Product
C. Intermediate
D. Catalyst

244. In the mechanism shown, NO_2 is a(n):

A. Reactant
B. Product
C. Intermediate
D. Catalyst

245. In the mechanism shown, O is a(n):

A. Reactant
B. Product
C. Intermediate
D. Catalyst

246. Which of the following shows the correct net reaction for the mechanism above?

A. $NO \rightarrow NO_2$
B. $O_3 \rightarrow O_2$
C. $O_3 + O \rightarrow 2O_2$
D. $NO + O_3 \rightarrow NO + O_2$

---

247. Scientist A proposes a mechanism for a certain reaction, and uses that mechanism to derive a rate law for the reaction. Scientist B then determines the rate law for the reaction experimentally, using the method of initial rates. If the two rate laws are the same, what can be concluded?

A. Assuming that the mechanism proposed by scientist A is correct, scientist B determined the correct experimental rate law.
B. Assuming that scientist B determined the correct experimental rate law, then the mechanism proposed by scientist A is correct.
C. Both the mechanism proposed by scientist A and the experimental rate law determined by scientist B are correct.
D. Either scientist A or scientist B made a mistake: rate laws determined by different techniques should be different.

248. The rate of a reaction between a solid and a liquid depends on:

I. The temperature of the liquid
II. The surface area of the solid that is in contact with the liquid
III. The presence of a catalyst

A. I only
B. III only
C. I and III only
D. I, II, and III

249. The rate constant for a reaction depends on which of the following?

I. Temperature
II. Concentration of reactants
III. Concentration of products

A. I only
B. I and II only
C. II and III only
D. I, II, and III

250. Which of the following can be changed by a catalyst?

I. The rate constant of a reaction
II. The rate law of a reaction
III. The equilibrium constant of a reaction

A. I only
B. I and II only
C. I and III only
D. I, II, and III

251. For all reactants at 1 *M* concentration, a certain reaction proceeds at an initial rate of 0.030 Ms^{-1}. The reaction is then repeated under identical conditions, except that the temperature is doubled. What is the new initial rate?

A. 0.015 Ms^{-1}
B. 0.030 Ms^{-1}
C. 0.060 Ms^{-1}
D. Cannot be determined from the information given

252. The reaction:

$$N_2(g) + 3H_2(g) \rightarrow 2NH_3(g)$$

proceeds very slowly at room temperature, but at a greater rate at higher temperatures. This is because:

A. The reaction is exothermic.
B. The reaction is endothermic.
C. At higher temperatures, collisions occur more frequently and are more likely to have sufficient energy to initiate the reaction.
D. At higher temperatures, the equilibrium constant is greater.

253. Suppose a scientist makes the claim that, when two reactions are in competition, the one that occurs at the greater rate always dominates. What criticism could be made of this scientist's claim?

A. The reaction that occurs at the greater rate also depletes its reactants faster, so the slower reaction will eventually surpass it.
B. A reaction that dominates at one temperature may not dominate an another temperature.
C. The scientist has ignored the role of the reverse reactions.
D. Reactions are never in competition.

254. Which of the following is a true statement?

A. The rate constant of a given reaction is independent of temperature.
B. Catalysts never change the mechanism of a reaction.
C. The rates of formation of different products in the same reaction under the same conditions are always the same.
D. Most collisions between reactants do not result in products being formed.

Use the following information to answer questions 255-257.

Using calculus, the rate law for a first order reaction can be transformed to:

$$\log[A]_t = -kt/2.303 + \log[A]_0$$

where $[A]_t$ is the concentration of A at time *t*, $[A]_0$ is the initial concentration of A, and *k* is the rate constant of the reaction. Methyl isonitrile is converted to acetonitrile in a first order process as follows:

$$CH_3NC(g) \rightarrow CH_3CN(g)$$

255. Which of the following graphs best illustrates the partial pressure of methyl isonitrile as the reaction goes forward?

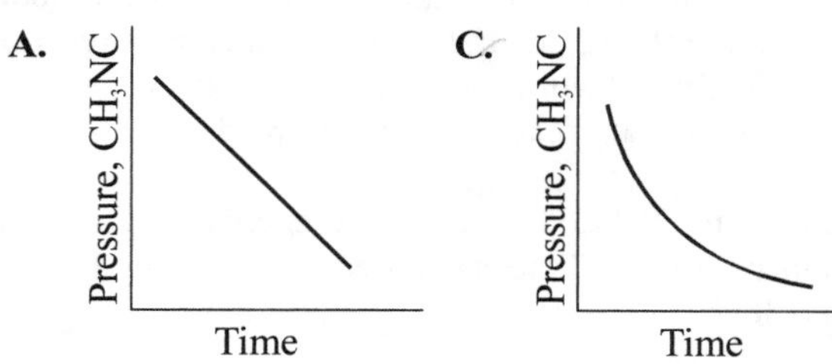

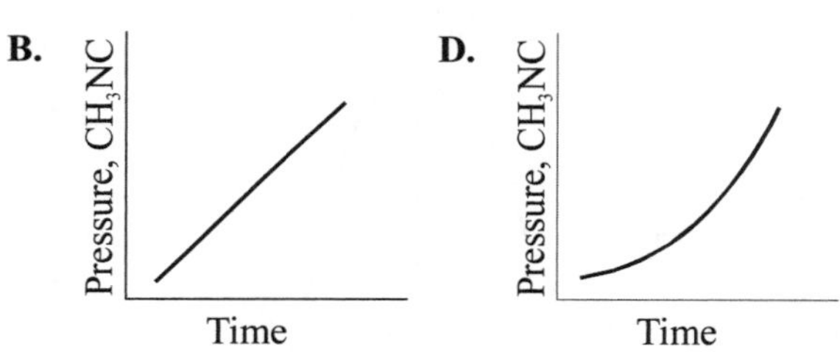

256. Why can pressure be used as a unit of concentration for a gas in the above reaction?

A. The pressure is directly proportional to the number of moles per unit volume.
B. The pressure is inversely proportional to the number of moles per unit volume.
C. The pressure and concentration have the same value.
D. The pressure of a gas is easier to measure than the concentration of a gas.

257. The rate constant for the conversion of methyl isonitrile is 5×10^{-5} s^{-1}. A scientist has a container containing methyl isonitrile gas with a partial pressure of 100 torr. After 12.8 hours (approx. 46,000 seconds), what is the partial pressure of methyl isonitrile gas inside the container?

A. 1 torr
B. 10 torr
C. 25 torr
D. 50 torr

258. Which of the following is NOT a difference between a first order elementary reaction and a second order elementary reaction?

A. The half life of a first order elementary reaction is independent of the starting concentration of the reactant, while the half life of a second order elementary reaction is dependent on the starting concentration of the reactant.
B. A first order elementary reaction is unimolecular, while a second order elementary reaction is bimolecular.
C. The rate of a first order elementary reaction is greater than the rate of a second order elementary reaction because no collision is required for the first order reaction.
D. When the concentrations of reactants are doubled, the rate of a first order elementary reaction is doubled while the rate of a second order elementary reaction is quadrupled.

For questions 259-263, refer to the energy diagram for the reaction pathway of the isomerization of methyl isonitrile given below.

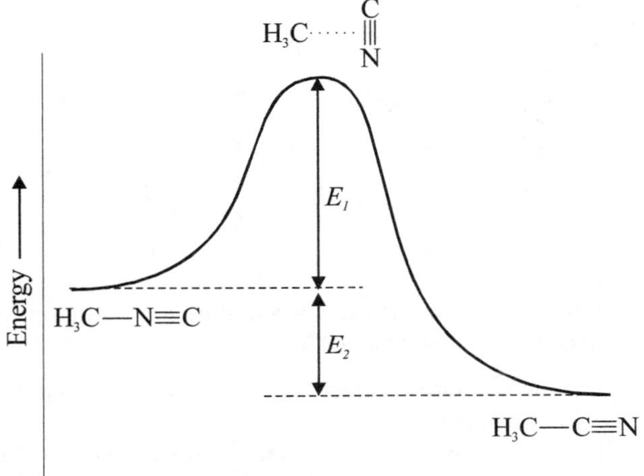

259. In the reaction pathway shown above, what is the the proper term to describe the following?

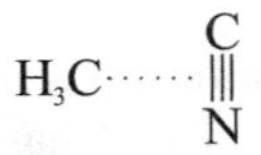

$H_3C\cdots C\equiv N$

A. intermediate
B. activated complex
C. catalyst
D. ionic complex

260. In the reaction pathway shown above, which of the following represents the energy of activation for the forward reaction?

A. E_1
B. E_2
C. E_1 plus E_2
D. E_2 minus E_1

261. In the reaction pathway shown above, which of the following represents the energy of activation for the reverse reaction?

A. E_1
B. E_2
C. E_1 plus E_2
D. E_2 minus E_1

262. Which of the following is true concerning the isomerization of methyl isonitrile?

A. Energy is absorbed to begin the reaction, and energy is absorbed by the overall reaction.
B. Energy is absorbed to begin the reaction, but energy is released by the overall reaction.
C. Energy is released to begin the reaction, but energy is absorbed by the overall reaction.
D. Energy is released to begin the reaction, but energy is released by the overall reaction.

263. If a catalyst were added to this reaction which of the following would decrease?

A. E_1 only
B. E_2 only
C. E_1 and E_2
D. neither E_2 or E_1

Use the energy diagram below to answer questions 264-266. The energy diagram below compares the kinetic energy of molecules in a reaction with the fraction of molecules colliding in that reaction. T_1 and T_2 represent the same reaction at different temperatures.

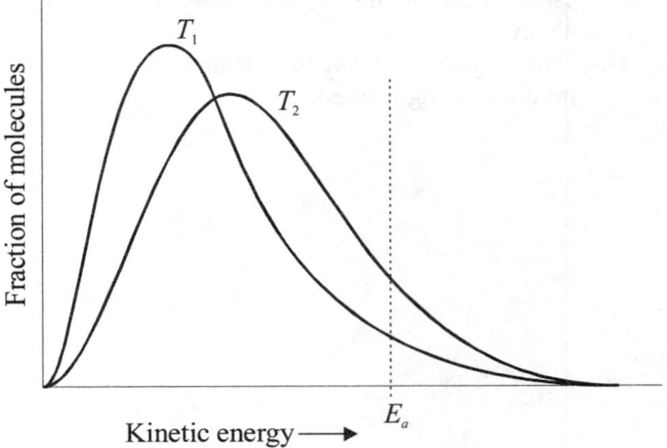

264. Which of the following is true concerning the temperatures T_1 and T_2?

- **A.** T_1 is greater than T_2 because the greatest number of collisions occur at a single energy level under T_1.
- **B.** T_1 is greater than T_2 because more collisions occur at higher energies under T_1.
- **C.** T_2 is greater than T_1 because the greatest number of collisions occur at a single energy level under T_2.
- **D.** T_2 is greater than T_1 because more collisions occur at higher energies under T_2.

265. Which of the following is true concerning the reaction at temperature T_1?

- **A.** A molecular collision occurring at any energy level *will* result in a reaction.
- **B.** A molecular collision occurring at any energy level *might* result in a reaction.
- **C.** A molecular collision occurring at any energy level above the energy of activation *will* result in a reaction.
- **D.** A molecular collision occurring at any energy level above the energy of activation *might* result in a reaction.

266. What changes to the diagram would reflect the addition of a catalyst to the reaction?

- **A.** The curves would be flattened and spread over a broader kinetic energy range.
- **B.** The curves would be heightened and narrowed.
- **C.** The curves would be shifted to the right.
- **D.** The vertical line marking the energy of activation would be shifted to the left.

---

267. According to the collision model, what three conditions are required for a reaction to occur?

- **A.** 1) a molecular collision; 2) sufficient temperature; 3) proper spatial orientation of the molecules
- **B.** 1) a molecular collision; 2) sufficient temperature; 3) sufficient duration of molecular contact
- **C.** 1) a molecular collision; 2) sufficient energy of collision; 3) proper spatial orientation of the molecules
- **D.** 1) a molecular collision; 2) sufficient energy of collision; 3) sufficient duration of molecular contact

The following are the elementary steps of a reaction. Refer to these steps to answer questions 268-273. The *k*s represent the rate constants for the respective reactions.

Step 1: $NO(g) + Br_2(g) \underset{k_{-1}}{\overset{k_1}{\rightleftharpoons}} NOBr_2(g)$ (fast)

Step 2: $NOBr_2(g) + NO(g) \xrightarrow{k_2} 2NOBr(g)$ (slow)

268. What is the overall reaction?

- **A.** $2NO(g) + Br_2(g) \rightarrow 2NOBr(g)$
- **B.** $NO(g) + Br\,(g) \rightarrow NOBr(g)$
- **C.** $NO(g) + Br_2(g) \rightarrow NOBr_2(g)$
- **D.** $NOBr_2(g) + Br_2(g) \rightarrow 2NOBr(g)$

269. Which of the following is true concerning the reaction?

- **A.** Step 1 is the rate determining step. The reaction will proceed at the rate of Step 1.
- **B.** Step 2 is the rate determining step. The reaction will proceed at the rate of Step 2.
- **C.** Step 1 is the rate determining step. The reaction will proceed at the rate of Step 1 plus Step 2.
- **D.** Step 2 is the rate determining step. The reaction will proceed at the rate of Step 1 plus Step 2.

270. Which of the following expressions gives the rate of the reaction?

- **A.** $k_1[NO][Br_2]$
- **B.** $k_2[NOBr_2][NO]$
- **C.** $(k_1 + k_2)[NO]^2[Br_2]$
- **D.** $k_2[NOBr_2][NO]^2[Br_2]$

271. Because $NOBr_2$ is a product of the fast step and a reactant of the slow step, and because the fast step precedes the slow step, which of the following concentrations of $NOBr_2$ should be used in the rate law for Step 2?

- **A.** zero moles per liter
- **B.** The concentration of $NOBr_2$ that exists when Step 1 has run to completion
- **C.** The equilibrium concentration of $NOBr_2$
- **D.** The concentration for $NOBr_2$ in Step 2 cannot be predicted because Step 1 is the fast step.

272. What is the equilibrium constant K for Step 1 in terms of the rate constant *k*s?

- **A.** k_1
- **B.** k_{-1}
- **C.** $k_1 + k_{-1}$
- **D.** k_1/k_{-1}

273. Which of the following expressions gives the rate constant k_c for the rate law of the complete reaction?

A. $k_2k_1k_{-1}$
B. k_2k_1/k_{-1}
C. k_2k_{-1}/k_1
D. $k_2 + k_1 + k_{-1}$

Equilibrium

274. Consider the following reaction:

$NH_3(aq) + HC_2H_3O_2(aq) \rightarrow NH_4^+(aq) + C_2H_3O_2^-(aq)$

This reaction is found to stop before the reactants are completely converted to products. What is a possible explanation?

A. As the quantity of reactants decreases, ammonia molecules and acetic acid molecules stop colliding.
B. As the quantity of products increases, the acetic acid begins to dissociate, stopping the reaction.
C. The reverse rate increases and the forward rate decreases until they exactly balance.
D. The catalyst is used up.

275. Reaction chambers A and B both contain equal amounts of pure 1-butene. A catalyst for the isomerization of 1-butene to 2-butene is added to chamber A. Assuming there are no other products, and that both chambers eventually reach equilibrium:

A. chamber A will always contain at least as much 2-butene as chamber B, with the difference increasing over time.
B. chamber A will always contain at least as much 2-butene as chamber B, with the difference first increasing, but then decreasing, over time.
C. chamber A will always contain at least as much 2-butene as chamber B, with the difference first decreasing, but then increasing, over time.
D. chamber B will always contain at least as much 2-butene as chamber A, with the difference increasing over time.

276. Consider the reaction:

$$4NH_3(g) + 5O_2(g) \rightarrow 4NO(g) + 6H_2O(g)$$

What is the equilibrium expression?

A. $\dfrac{[H_2O]^6[NO]^4}{[NH_3]^4[O_2]^5}$

B. $[H_2O]^6[NO]^4$

C. $\dfrac{[NO]^4}{[NH_3]^4[O_2]^5}$

D. Cannot be determined from the information given.

277. Consider the reaction:

$$2NaCl(aq) + 2H_2O(l) \rightarrow 2NaOH(aq) + Cl_2(g) + H_2(g)$$

What is the equilibrium expression?

A. $\dfrac{[NaOH]^2[Cl_2][H_2]}{[NaCl]^2[H_2O]^2}$

B. $[NaOH]^2[Cl_2][H_2]$

C. $\dfrac{[NaOH]^2[Cl_2][H_2]}{[NaCl]^2}$

D. $[Cl_2][H_2]$

278. Consider the following conversion

$$O_2(g) \rightarrow O_2(aq)$$

What is the equilibrium expression?

A. $[O_2(aq)]/[O_2(g)]$
B. $[O_2(aq)]$
C. $1/[O_2(g)]$
D. $[O_2(g)]$

279. Consider the reaction:

$$ZnO(s) + CO(g) \rightarrow Zn(s) + CO_2(g)$$

What is the equilibrium expression?

A. $K = \dfrac{[Zn][CO_2]}{[ZnO][CO]}$

B. $K = [Zn][CO_2]$

C. $K = \dfrac{[ZnO][CO]}{[Zn][CO_2]}$

D. $K = \dfrac{[CO_2]}{[CO]}$

280. Consider the reaction:

$$[Fe(H_2O)_6]^{2+}(aq) + Cl^-(aq) \rightarrow [FeCl(H_2O)_5]^+(aq) + H_2O(l)$$

What is the equilibrium expression?

A. $K = [FeCl(H_2O)_5^+][H_2O]/[Fe(H_2O)_6^{2+}][Cl^-]$
B. $K = [FeCl(H_2O)_5^+][H_2O]$
C. $K = [Fe(H_2O)_6^{2+}][Cl^-]/[FeCl(H_2O)_5^{2+}][H_2O]$
D. $K = [FeCl(H_2O)_5^+]/[Fe(H_2O)_6^{2+}][Cl^-]$

281. Consider the following reaction:

$$2NO_2(g) \rightarrow 2NO(g) + O_2(g)$$

Suppose it is found that, at equilibrium, the concentrations of nitrogen dioxide, nitrogen monoxide, and oxygen were 5.0 *M*, 0.10 *M*, and 0.15 *M* respectively. What is the equilibrium constant for this reaction?

A. 6×10^{-5}
B. 3×10^{-3}
C. 0.25
D. 5.0

282. Suppose the reaction:

$$CO(g) + Cl_2(g) \rightarrow COCl_2(g)$$

has an equilibrium constant of 530. What is the equilibrium constant of the reaction:

$$2CO(g) + 2Cl_2(g) \rightarrow 2COCl_2(g)$$

A. 530
B. 1060
C. 2120
D. 2.81×10^5

283. Consider the following reaction:

$$H_2(g) + I_2(g) \rightarrow 2HI(g)$$

At 740 K, this reaction has an equilibrium constant of 50. If, at 740 K, the concentration of hydrogen gas is 0.2 M, iodine gas is 0.3 M, and hydrogen iodide gas is 5 M:

A. the concentration of hydrogen iodide will increase over time.
B. the concentration of hydrogen iodide will decrease over time.
C. the system is at equilibrium.
D. This combination of concentrations is inconsistent with the information given

284. Consider the reaction:

$$H_2(g) + I_2(g) \rightarrow 2HI(g)$$

At 740 K, this reaction has an equilibrium constant of 50. If, at 740 K, the concentration of hydrogen gas is 2 M, iodine gas is 3 M, and hydrogen iodide gas is 6 M:

A. the concentration of hydrogen iodide will increase over time.
B. the concentration of hydrogen iodide will decrease over time.
C. the system is at equilibrium.
D. This combination of concentrations is inconsistent with the information given

285. In considering the reaction:

$$3NO_2(g) + H_2O(l) \rightarrow 2HNO_3(aq) + NO(g)$$

student A writes the equilibrium expression as:

$$K = \frac{[NO][HNO_3]^2}{[NO_2]^3}$$

while student B writes:

$$K = \frac{P_{NO}[HNO_3]^2}{[NO_2]^3}$$

Which of the following is a true statement?

A. Student B is incorrect; it is not permissible to mix concentrations and partial pressures in one equilibrium expression.
B. Student A is incorrect; gases should always appear as partial pressures in equilibrium expressions.
C. Both students are correct, and will arrive at the same value of K.
D. Both students are correct, but they will arrive at different values of K.

286. Consider the following reaction:

$$2H_2(g) + O_2(g) \rightarrow 2H_2O(g)$$

Which of the following would decrease the proportion of water vapor at equilibrium?

A. Increasing the partial pressure of oxygen gas
B. Decreasing the volume of the container
C. Reducing the amount of catalyst present
D. Raising the temperature

287. Consider the following endothermic reaction:

$$NH_4HS(s) \rightleftharpoons NH_3(g) + H_2S(g)$$

If the system were at equilibrium within a sealed container, which of the following would produce additional H_2S?

I. Raising the temperature
II. Adding more NH_4HS
III. Increasing the volume of the container

A. II only
B. I and III only
C. II and III only
D. I, II, and III

288. Consider the following reaction at equilibrium:

$$2H_2S(g) + SO_2(g) \rightarrow 3S(s) + 2H_2O(l)$$

Which of the following changes would shift the equilibrium to the right?

I. Increasing the total pressure by increasing the partial pressure of hydrogen sulfide gas
II. Increasing the total pressure by adding helium
III. Increasing the total pressure by decreasing the volume of the container

A. I only
B. I and II only
C. I and III only
D. I, II, and III

289. Consider the following exothermic reaction and LeChatelier's principle:

$$2CO(g) + O_2(g) \rightarrow 2CO_2(g)$$

If a container holds only CO and O_2, what effect will raising the temperature have on the forward reaction?

A. The forward reaction rate will increase.
B. The forward reaction rate will remain the same.
C. The forward reaction rate will decrease.
D. The reaction will not go forward.

For Questions 290-295 refer to the Haber Process shown below:

$$N_2(g) + 3H_2(g) \rightleftharpoons 2NH_3(g)$$

In 1912 Fritz Haber developed the Haber process for making ammonia from nitrogen and hydrogen. His development was crucial for the German war effort of World War I, providing the Germans with ample fixed nitrogen for the manufacture of explosives.

The Haber process takes place at 500 °C and 200 atm. It is an exothermic reaction. The graph below shows the change in concentrations of reactants and products as the reaction progresses.

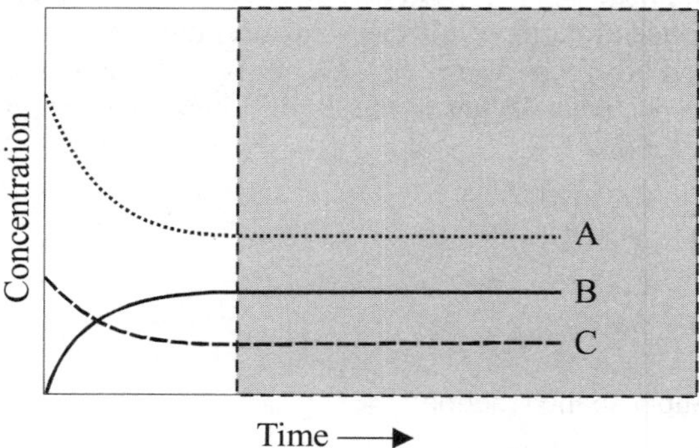

290. Even at the high temperatures, the conversion of nitrogen and hydrogen to ammonia was slow. In order to make the process industrially efficient Fritz Haber used a metal oxide catalyst. Which of the following was accomplished by use of the metal oxide catalyst?

I. The rate of production of ammonia increased.
II. The energy of activation was raised.
III. The equilibrium shifted to the right.

A. I only
B. II only
C. I and II only
D. I, II, and III

291. Instead of developing a catalyst, why didn't Haber increase the rate of the reaction by raising the temperature?

A. The Haber process is exothermic, so raising the temperature would have lowered the rate of the reaction.
B. The Haber process is exothermic, so raising the temperature would have reduced the yield of the reaction.
C. Higher temperatures would have caused an increase in pressure lowering the yield of the reaction.
D. Higher temperatures might have decomposed the hydrogen.

292. In the diagram, the letters A, B, and C represent which of the following?

A. A = NH_3, B = H_2, and C = N_2
B. A = NH_3, B = N_2, and C = H_2
C. A = H_2, B = NH_3, and C = N_2
D. A = N_2, B = NH_3, and C = H_2

293. Which of the following most accurately describes the shaded area of the graph?

A. A limiting reagent has been used up.
B. The catalyst has been used up.
C. The reaction has reached equilibrium.
D. The reaction has run to completion.

294. Which of the following graphs could also accurately reflect the establishment of equilibrium between nitrogen, hydrogen, and ammonia?

A.

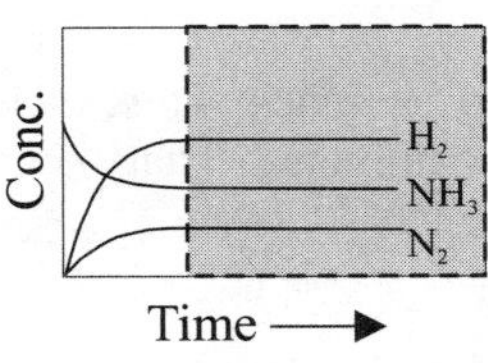

C.

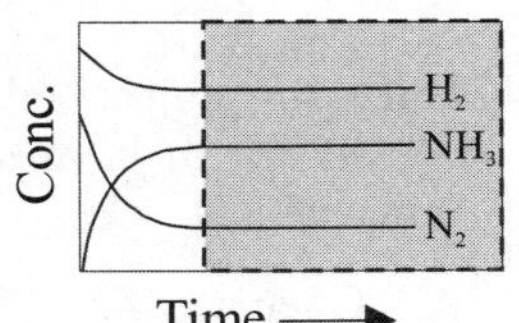

B.

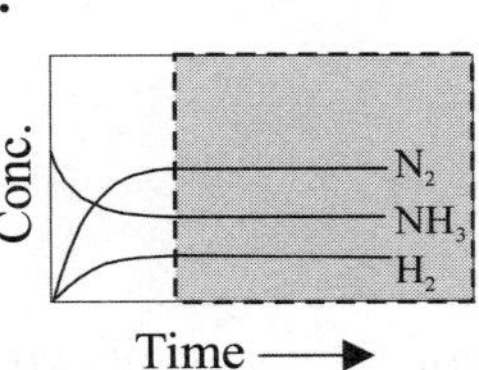

D.

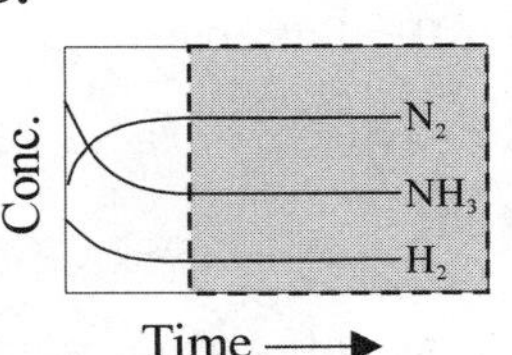

295. Assume that no catalyst was used to arrive at the results in the graph. If the same reaction were run in the presence of a catalyst, which of the following graphs would most accurately represent the results?

A.

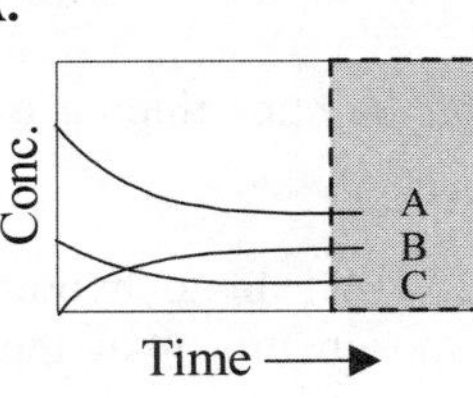

C.

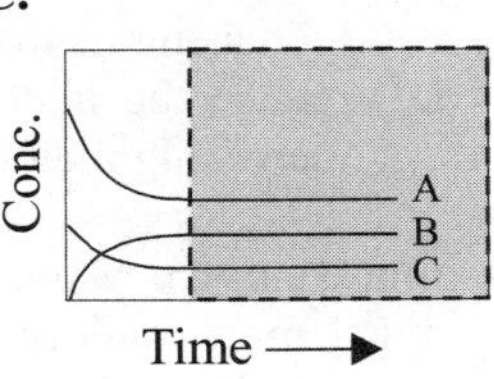

B.

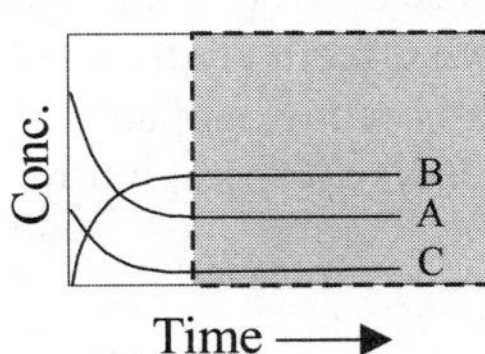

D.

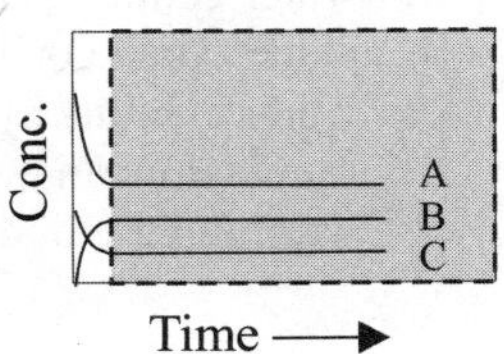

Systems

296. In an isolated system, which of the following can be exchanged between the system and its surroundings?

A. Energy only
B. Matter only
C. Both matter and energy
D. Neither matter nor energy

297. In a closed system, what can be exchanged between the system and its surroundings?

A. Energy only
B. Matter only
C. Both matter and energy
D. Neither matter nor energy

298. In an open system, what can be exchanged between the system and its surroundings?

A. Energy only
B. Matter only
C. Both matter and energy
D. Neither matter nor energy

299. A person pushes down on a leak-proof piston, compressing the gas inside. The gas is what kind of system?

A. Open
B. Closed
C. Isolated
D. Entropic

300. A person blows air into a balloon. The gas in the balloon is what kind of system?

A. Open
B. Closed
C. Isolated
D. Entropic

301. A person blows air through an open pipe, so that air enters the pipe at the same rate that it leaves. The air in the pipe is what kind of system?

A. Open
B. Closed
C. Isolated
D. Entropic

302. A glass of water is left under a bell jar, which is then covered by a layer of opaque, effective insulation. After some time, it is found that part of the water has evaporated. The contents of the bell jar are what kind of system?

A. Open
B. Closed
C. Isolated
D. Entropic

303. A grenade is exploded inside a sealed concrete bunker. The shielding of the bunker is such that no vibrations, concussion, or heat are felt by soldiers leaning against the exterior wall. The contents of the bunker are what kind of system?

A. Open
B. Closed
C. Isolated
D. Entropic

304. Once an object falls into a black hole, scientists consider it to have left our universe. That means our universe is:

A. Open
B. Closed
C. Isolated
D. Entropic

Heat

305. A "coffee-cup" calorimeter often consists of two nested styrofoam cups with a lid. The primary purpose of using two coffee cups is to:

A. reduce convection through the styrofoam by providing a thicker barrier
B. reduce conduction through the styrofoam by providing a thicker barrier
C. increase the heat capacity of the calorimeter by increasing its mass
D. reduce the likelihood of the cups tipping over by increasing their mass

306. Immersion in water at 5°C will lead to hypothermia much more rapidly than exposure to air at the same temperature because:

A. water conducts heat less effectively than air.
B. water conducts heat more effectively than air.
C. water exhibits hydrogen bonding, and air does not.
D. immersion in water prevents perspiration from warming the body.

307. A sealed container with hollow walls is quite effective at maintaining its inside temperature. What is the purpose of the hollow walls?

A. The air in the hollow walls is an excellent insulator, cutting down on conduction.
B. The air in the hollow walls provides an additional source of heat for the container.
C. The hollow walls trap air trying to escape from the box, reducing convection.
D. Reactions can take place within the walls, helping to maintain the temperature of the container.

308. Which of the following aspects of thermos design minimizes heat conduction?

A. A tight-fitting, screw-on lid
B. Double-walled construction
C. Heavy-duty aluminum construction
D. Shiny interior coating

309. What aspect of thermos design minimizes heat convection?

A. A tight-fitting, screw-on lid
B. Double-walled construction
C. Heavy-duty aluminum construction
D. Shiny interior coating

310. What aspect of thermos design minimizes heat radiation?

A. A tight-fitting, screw-on lid
B. Double-walled construction
C. Heavy-duty aluminum construction
D. Shiny interior coating

311. Madam Curie, when considering how to keep a cup of coffee warm, said that first she would put a lid on the cup. This indicates Madam Curie was most concerned about heat loss due to:

A. conduction.
B. convection.
C. radiation.
D. sublimation.

312. Which of the following properties of a person are state functions?

I. Height
II. Age
III. Cholesterol level

A. I only
B. III only
C. I and III only
D. I, II, and III

313. Which of the following properties of a gas are state functions?

I. Temperature
II. Heat
III. Work

A. I only
B. I and II only
C. II and III only
D. I, II, and III

314. The *Helmholz function*, *F*, is a state function. From this fact alone, what can be concluded about *F*?

A. In a closed system, it is conserved.
B. In an open system, it is not conserved.
C. It must be a function of pressure, volume, and temperature only.
D. If the system is taken from one state to another, the change in *F* will be independent of the process used.

315. Considering that all objects radiate heat, which of the following statements must be true?

A. All objects are continually getting colder.
B. Any object that gets warmer must be experiencing conduction or convection.
C. No object can reach a temperature of absolute zero.
D. The First Law of Thermodynamics does not apply when radiation is considered.

Use the following information to answer questions 316-318. Objects radiate heat at a rate given by the following equation:

$$P = \sigma\varepsilon AT^4$$

where P is the rate at which heat is radiated, σ is the Stefan-Boltzman constant ($5.67\text{x}10^{-8}$ W m^{-2} K^{-4}), ε is the emissivity of the object's surface having a value between zero and one, A is the surface area, and T is the temperature in Kelvin.

The rate at which objects absorb heat from their environment is given by the same equation except that T is the temperature of the environment. The net rate of energy exchange due to radiative processes is given by:

$$P_n = \sigma\varepsilon A(T_{env}^4 - T^4)$$

where P_n is the net rate of energy exchange and T_{env} is the temperature of the environment.

316. A black body radiator has an emissivity of:

A. 0
B. 0.5
C. 1.0
D. The emissivity of a blackbody radiator may vary.

317. Which of the following is true of a blackbody radiator?

A. It reflects all radiation that it intercepts.
B. It absorbs all radiation that it intercepts.
C. It does not radiate.
D. It does not absorb radiation.

318. An object is 10 degrees warmer than the room in which it sits. In which of the following circumstances will the object reach room temperature the fastest?

A. The room is 0 °C and the object is white.
B. The room is 0 °C and the object is black.
C. The room is 10 °C and the object is white.
D. The room is 10 °C and the object is black.

---

319. Indigenous inhabitants of desert regions often wear loose black robes instead of white robes. Which of the following is the best explanation?

A. Black robes stay cooler in the sun.
B. Black robes don't emit as much radiation and thus keep the wearer cool.
C. White robes reflect the heat into the wearer whereas black robes do not.
D. Black robes warm the air between the robe and the skin causing it to rise. The wearer is cooled by convection.

320. Which of the following requires the most energy?

A. Bringing a small pot of water to a boil slowly.
B. Bringing a small pot of water to a boil quickly.
C. Bringing a large pot of water to a boil slowly.
D. Bringing a large pot of water to a boil quickly.

321. Which of the following could be one reason why astronauts' uniforms are white?

A. In the cold dark of space, white reflects more than black.
B. In the cold dark of space, white doesn't radiate as much as black.
C. In the warm sunshine, white absorbs more than black.
D. In the warm sunshine, white doesn't radiate as much as black.

Use the following information to answer questions 322-327.
The rate of heat transfer through a slab is given by the equation:

$$\frac{Q}{t} = kA\frac{T_h - T_c}{L}$$

where Q is heat, t is time, k is the thermal conductivity constant, A is the cross-sectional area of each slab face, L is the length of the slab, and T_h and T_c are the hot and cold ends of the slab respectively.

Six contiguous slabs at six temperatures, T_1, T_2, etc.., are placed in a row as shown. Heat flows from left to right through the slabs. The cross-section of each slab face is square. (Unless otherwise stated assume that each slab is made from the same material and the temperature of the slabs do not change over time.)

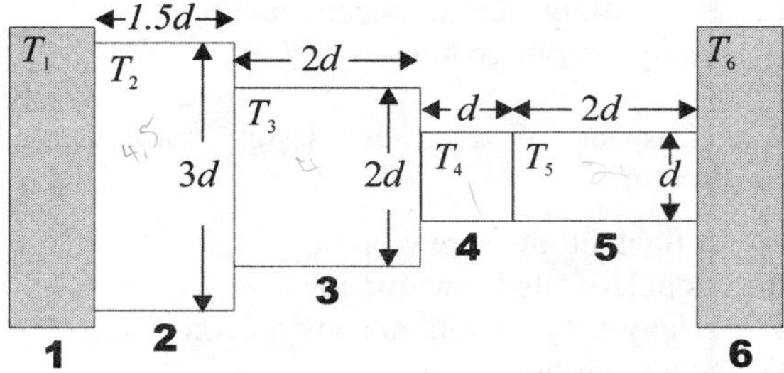

322. Which of the following gives the temperatures of the four middle slabs from greatest to smallest?

A. $T_2 > T_3 > T_4 > T_5$
B. $T_5 > T_4 > T_3 > T_2$
C. $T_3 > T_5 > T_2 > T_4$
D. $T_4 > T_2 > T_5 > T_3$

323. Through which slab is the rate of heat flow, Q/t, the greatest?

A. slab 2
B. slab 3
C. slab 4
D. The rate of heat flow is the same through all slabs.

324. Consider only slabs 2, 3, and 4. Which slab experiences the greatest temperature difference between its left and right sides?

A. slab 2
B. slab 3
C. slab 4
D. The temperature difference across each of these slabs is the same.

325. Which of the following would increase the rate of heat flow across slab 3?

A. decreasing both T_1 and T_6 by 25 °C
B. increasing both T_1 and T_6 by 25 °C
C. decreasing the length of slab 3
D. increasing the length of slab 3

326. If each slab is made of a different material and the temperature difference is the same across each slab, which slab has the greatest thermal conductivity constant, *k*?

A. slab 2
B. slab 3
C. slab 4
D. slab 5

327. The heat transfer through the slabs is due to:

A. convection
B. radiation
C. conduction
D. translation

---

328. Which of the following phases of a substance typically has the greatest resistance to conduction?

A. solid
B. liquid
C. gas
D. It depends upon the substance.

Work

329. The net work done on a certain gas is 55 J, and the net heat flow into the gas is –23 J. What is the net work done *by* the gas?

A. –55 J
B. 0 J
C. 32 J
D. 78 J

330. The net work done on a certain gas is 55 J, and the net heat flow into the gas is –23 J. What is the change in energy of the gas?

A. –55 J
B. 0 J
C. 32 J
D. 78 J

331. A heat engine is designed using an ideal gas as a working fluid. At the completion of each cycle, the engine returns to its original state. If the *net* flow of heat into the heat engine is 60,000 kJ/cycle, what can be said about the net work done each cycle? Neglect losses due to friction.

A. It is less than 60,000 kJ
B. It is equal to 60,000 kJ
C. It is greater than 60,000 kJ
D. It begins as 60,000 kJ, and decreases on each subsequent cycle

332. A gas is compressed from a volume of 5 m^3 to a volume of 3 m^3 by a constant pressure of 50,000 Pa. What is the work done on the gas by this process?

A. –500,000 J
B. –100,000 J
C. 100,000 J
D. 300,000 J

333. Fuel cells have been created that employ the following reaction to produce energy:

$$2H_2 + O_2 \rightarrow 2H_2O$$

These fuel cells produce a considerable amount of usable energy without creating any pollution, and they require only the plentiful elements hydrogen and oxygen. Oxygen is plentiful in our atmosphere, and hydrogen is readily produced by the electrolysis of water. Based on this description, are fuel cells a viable alternative to other energy sources, such as fossil fuel, nuclear power, or solar power?

A. Yes, because it is a highly exothermic reaction
B. Yes, because electrolyzing water to provide hydrogen would require less energy than the fuel cell would produce, since oxygen is so readily available
C. No, because the entropy of the reaction is unfavorable
D. No, because electrolyzing water to provide hydrogen will require as much energy as the fuel cell would produce

334. Which of these systems violate(s) the first law of thermodynamics?

I. A pendulum, perfectly isolated from its environment, which swings forever
II. A battery that never "dies" or needs recharging
III. A refrigerator that uses all of the heat it removes from its interior to provide electricity to an apartment

A. I only
B. II only
C. I and III only
D. I, II, and III

335. Which of the following systems violate(s) the second law of thermodynamics?

I. A pendulum, perfectly isolated from its environment, which swings forever
II. A battery that never "dies" or needs recharging
III. A refrigerator that uses all of the heat it removes from its interior to provide electricity to an apartment

A. I only
B. III only
C. I and II only
D. I, II, and III

Refer to the diagram below to answer questions 336-346. The diagram below displays the pressure and volume of a sample of ideal gas as it is taken very slowly through six states: A, B, C, D, E, and F, along the pathway shown. (R = 8.314 J k^{-1} mol^{-1})

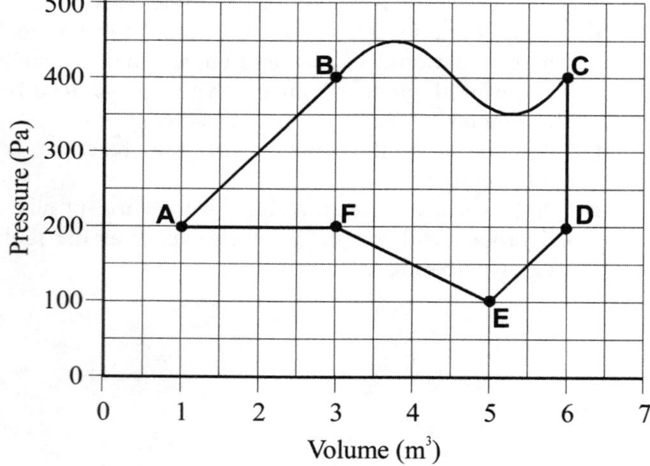

336. Approximately what is the net work done on the gas as it goes directly from state F to state A?

A. –400 J
B. 0 J
C. 400 J
D. 600 J

337. Approximately what is the net work done on the gas as it goes directly from state C to state D?

A. –1200 J
B. 0 J
C. 1200 J
D. 2400 J

338. Approximately what is the net work done on the gas as it goes directly from state A to state B?

A. –1000 J
B. –600 J
C. –400 J
D. 1000 J

339. Approximately what is the net work done on the gas as it goes directly from state B to state C?

A. –1200 J
B. 1200 J
C. 2400 J
D. An approximation cannot be made.

340. Approximately what is the net work done on the gas as it goes directly from state A to state B, and then directly back to state A?

A. –1200 J
B. –600 J
C. 0 J
D. 1200 J

341. Approximately what is the net work done on the gas as it goes from state A, through states B, C, D, E, and F, and back to state A?

A. –1600 J
B. –950 J
C. 0 J
D. 950 J

342. Approximately what is the net work done on the gas as it goes from state A, through states F, E, D, C, and B, and back to state A?

A. –1600 J
B. –950 J
C. 0 J
D. 950 J

343. As the gas is taken from A directly to F, the temperature of the gas:

A. increases
B. decreases
C. remains constant
D. depends upon the heat transfer

344. As the gas is taken from E directly to D, heat flow is:

A. into the gas.
B. out of the gas.
C. zero.
D. The direction of heat flow cannot be determined from the information given.

345. If the sample contains one mole of gas, approximately what is the change in temperature of the gas as it goes directly from state A to state B?

A. 120 K decrease
B. 24 K decrease
C. 120 K increase
D. The temperature must be constant.

346. If the sample contains one mole of gas, approximately what is the net heat flow into the gas as it goes directly from state A to state B?

A. –200 J
B. 0 J
C. 200 J
D. The net heat flow depends upon the specific heat of the gas.

Use the following information and diagram to answer questions 347-348. The diagram below shows an apparatus prepared to demonstrate the free adiabatic expansion of a gas. When the stopcock is opened, the gas will expand into the second chamber. The surrounding insulation will prevent any heat from escaping.

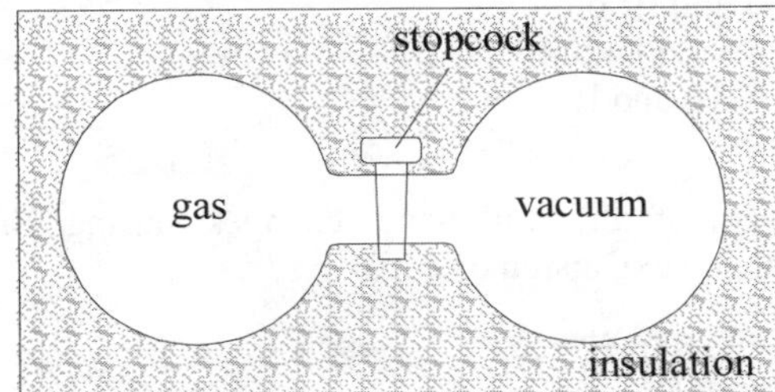

347. The work done in a free adiabatic expansion of a gas:

A. is zero.
B. is equal to the pressure times the change in volume.
C. is done only on the gas.
D. cannot be calculated because pressure changes as the expansion occurs.

348. The temperature of an ideal gas undergoing a free adiabatic expansion should:

A. remain constant.
B. increase because the molecules will be move faster as they escape into the vacuum.
C. decrease because the molecular energy is spread over a larger area.
D. decrease because the pressure goes down.

Thermodynamic Functions

349. Which of the following is/are intensive properties?

I. Volume
II. Density
III. Specific Heat

A. III only
B. I and II only
C. II and III only
D. I, II, and III

350. The *Helmholz function*, *F*, is an extensive property. If the change in *F* for the decomposition of 120 grams of a substance is −50 kJ, what is the change in *F* when 360 grams of the same substance decompose?

A. –17 kJ
B. –50 kJ
C. –150 kJ
D. –450 kJ

351. The *chemical potential*, μ, is an intensive property. If the change in μ for the decomposition of 120 grams of a substance is –50 kJ/mol, what is the change in μ when 360 grams of the same substance decompose?

A. –17 kJ
B. –50 kJ
C. –150 kJ
D. –450 kJ

Internal Energy

352. Which of the following are forms of internal energy?

I. Bond energy
II. Thermal energy
III. Gravitational energy

A. II only
B. I and II only
C. II and III only
D. None of these

353. For an ideal gas, internal energy depends upon:

I. pressure
II. temperature
III. volume

A. I only
B. II only
C. I and III only
D. I, II, and III

354. In which of the following cases does the internal energy of the described system increase?

I. A clay ball is dropped and sticks to the ground
II. Hydrogen and oxygen react exothermically to form water
III. A car is driven at a constant speed (consider the effect of air resistance)

A. II only
B. I and II only
C. I and III only
D. I, II, and III

355. When an exothermic reaction takes place, where does the energy which is released come from?

A. The surroundings
B. The kinetic energy of the reacting molecules
C. The potential energy of the reacting molecules
D. The thermal energy of the reactants

Temperature

356. If T_1 and T_2 are two temperatures, which of the following expressions will yield the same value whether both temperatures are given in Celsius or both are given in Kelvin?

A. $T_1 + T_2$
B. $T_1 - T_2$
C. T_1T_2
D. T_1/T_2

357. It takes 35 calories to raise 5 grams of a material by 10°C. How much heat is required to raise the 5 grams of material by 10 K?

A. 0.0037 cal
B. 35 cal
C. 308 cal
D. 9600 cal

358. A stationary gas has a kinetic energy of 500 J at 30°C. What is its kinetic energy at 60°C?

A. 250 J
B. 500 J
C. 550 J
D. 1000J

359. When the temperature of a bridge changes from –5°C on a winter night to 10°C in the afternoon, an iron bolt expands in diameter from 1.00000 cm to 1.00018 cm. What is the coefficient of linear expansion for iron? (Use the following formula: $\Delta L/L = \alpha\Delta T$.)

A. $(1.8 \times 10^{-4})(15)$
B. $(1.8 \times 10^{-4}) / 15$
C. $15 / (1.8 \times 10^{-4})$
D. $1 / (15 \times 1.8 \times 10^{-4})$

Use the following information and diagram to answer questions 360-361. The diagram below shows an apparatus prepared to demonstrate the free adiabatic expansion of a gas. When the stopcock is opened, the gas will expand into the second chamber. The surrounding insulation will prevent any heat from escaping.

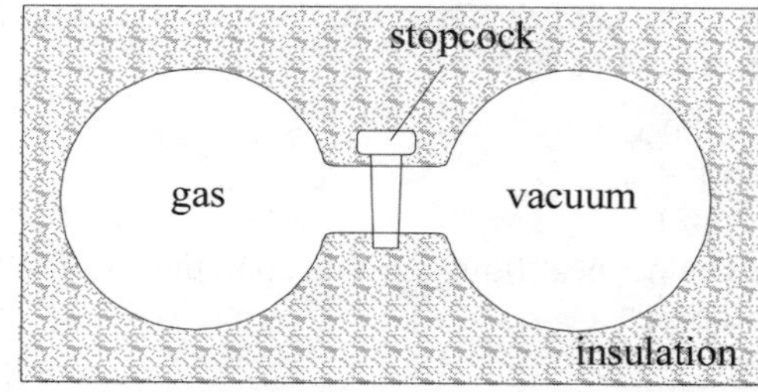

360. Which of the following changes during the free adiabatic expansion of an ideal gas?

I. internal energy
II. temperature
III. pressure

A. III only
B. I and III only
C. II and III only
D. I, II, and III

361. Which of the following changes during the free adiabatic expansion of a real gas?

I. internal energy
II. temperature
III. pressure

A. III only
B. I and III only
C. II and III only
D. I, II, and III

362. Two blocks are placed side by side so they are touching each other. Heat will flow from the first block to the second block if the first block has greater:

I. internal energy
II. temperature
III. density

A. I only
B. II only
C. III only
D. I and II only

Use the diagram below to answer questions 363-366.

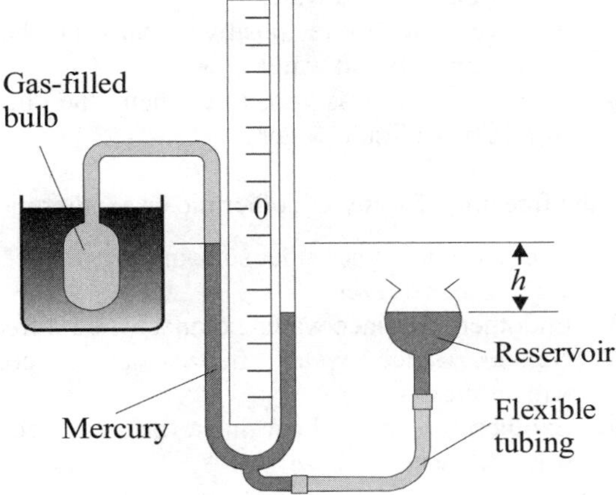

The diagram shows a gas thermometer. A gas filled bulb is connected to a mercury manometer. The reservoir can be raised or lowered so that the mercury level on the left is brought to the zero on the manometer. When this is done, the temperature T of any body in thermal equilibrium with the bulb is given as:

$$T = Cp$$

where p is the pressure of the gas and C is a constant. The pressure p is given by:

$$p = p_o - \rho gh$$

where p_o is atmospheric pressure and ρ is the density of mercury.

363. Why must the reservoir level match the level of the manometer?

A. Matching the level of the reservoir with the level of the manometer ensures that the gas remains at atmospheric pressure.
B. Matching the level of the reservoir with the level of the manometer ensures that the gas remains at constant temperature.
C. Matching the level of the reservoir with the level of the manometer ensures that the gas remains at constant pressure.
D. Matching the level of the reservoir with the level of the manometer ensures that the gas remains at constant volume.

364. Which of the following is true concerning the pressure of the gas in the diagram?

A. The gas is at atmospheric pressure.
B. The gas is below atmospheric pressure.
C. The gas is above atmospheric pressure.
D. The pressure of the gas relative to atmospheric pressure can't be determined until the mercury level is brought to zero on the manometer.

365. A student calculated the gas pressure with the reservoir positioned as shown in the diagram. He then used that gas pressure in the given equations to calculate the temperature of the fluid around the bulb. The temperature calculated by the student will be:

A. the correct temperature of the fluid in kelvins.
B. the correct temperature of the fluid in degrees Celsius.
C. less than the actual temperature of the fluid.
D. greater than the actual temperature of the fluid.

366. The thermometer works best when the gas behaves likes an ideal gas. Which of the following is most likely to increase the accuracy of the thermometer?

A. Using less gas in the bulb
B. Using more gas in the bulb
C. Using a heavier fluid in the tube
D. Using a lighter fluid in the tube

Enthalpy

367. Which of the following expressions defines enthalpy? (q is heat; U is internal energy; P is pressure, V is volume)

A. q
B. ΔU
C. $U + PV$
D. $U + q$

368. For an ideal gas, enthalpy depends upon:

I. pressure
II. temperature
III. volume

A. I only
B. II only
C. III only
D. I and III only

369. Under standard conditions, which of the following reactions will have the greatest difference between the enthalpy and energy of reaction?

A. $2C(graphite) + O_2(g) \rightarrow 2CO(g)$
B. $C(graphite) + O_2(g) \rightarrow CO_2(g)$
C. $C(graphite) \rightarrow C(diamond)$
D. $CO(g) + NO_2(g) \rightarrow CO_2(g) + NO(g)$

Questions 370-372 refer to the following reaction:

$$P_4(s) + 6Cl_2(g) \rightarrow 4PCl_3(l) \quad \Delta H° = -1279 \text{ kJ/mol}$$

370. The reaction is:

A. Endothermic
B. Exothermic
C. Spontaneous
D. Nonspontaneous

371. What is the standard enthalpy change for the following reaction?

$$3P_4(s) + 18Cl_2(g) \rightarrow 12PCl_3(l)$$

A. –3837 kJ/mol
B. –1279 kJ/mol
C. –426 kJ/ mol
D. 1279 kJ/mol

372. What is the standard enthalpy change for the following reaction?

$$4PCl_3(l) \rightarrow P_4(s) + 6Cl_2(g)$$

A. –3837 kJ/mol
B. –1279 kJ/mol
C. –426 kJ/ mol
D. 1279 kJ/mol

---

373. Given that the $\Delta H°$ of the following reaction is 92.2 kJ/mol, what is the heat of formation of ammonia gas?

$$2NH_3(g) \rightarrow N_2(g) + 3H_2(g)$$

A. –92.2 kJ/mol
B. –46.1 kJ/mol
C. 46.1 kJ/mol
D. 92.2 kJ/mol

374. The heat of formation of water vapor is:

A. Negative, and more negative than the heat of formation of liquid water
B. Negative, but less negative than the heat of formation of liquid water
C. Positive, and more positive than the heat of formation of liquid water
D. Positive, but less positive than the heat of formation of liquid water

375. Is the freezing of water endothermic or exothermic?

A. Endothermic, since the surroundings cool down when water freezes
B. Endothermic, since water expands when it freezes
C. Neither, since water freezes at a constant temperature
D. Exothermic, since heat must be removed from water to make it freeze

376. Which calculation would provide an estimate of the bond energy of an oxygen-oxygen double bond?

A. Finding the heat of formation of $O_2(g)$
B. Doubling the heat of formation of $O(g)$
C. Dividing the heat of formation of $CO_2(g)$ by two
D. Doubling the bond energy of an oxygen-oxygen single bond

377. Given that the bond energy of hydrogen-hydrogen bonds is 436 kJ/mol, that of hydrogen-oxygen bonds is 464 kJ/mol, and those in oxygen molecules 496 kJ/mol, what is the approximate heat of reaction for $2H_2 + O_2 \rightarrow 2H_2O$?

A. –488 kJ/mol
B. –440 kJ/mol
C. 440 kJ/mol
D. 488 kJ/mol

378. Which of the following reactions is likely to give off the most heat per *gram* of reactants?

A. $2H \rightarrow H_2$
B. $2O \rightarrow O_2$
C. $3O \rightarrow O_3$
D. $Xe + 3F_2 \rightarrow XeF_6$

379. The standard enthalpy of formation of oxygen gas is:

A. Negative, since bond formation is always exothermic
B. Zero, since oxygen gas is the most stable form of oxygen under standard conditions
C. Positive, since gas formation is always endothermic
D. Negative, zero, or positive depending on the temperature

380. The standard enthalpy of formation of atomic oxygen is:

A. Negative, since all spontaneous reactions are exothermic
B. Zero, since oxygen is an element
C. Positive, since breaking bonds is always endothermic
D. Negative, zero, or positive depending on the temperature

381. A single-step reaction has an activation energy of 25 kJ/mol and a heat of reaction of –85 kJ/mol. What is the activation energy of the reverse reaction?

A. –25 kJ/mol
B. 25 kJ/mol
C. 60 kJ/mol
D. 110 kJ/mol

382. Consider the following mechanism:

$$Br_2 + H_2O_2 \rightarrow 2Br^- + 2H^+ + O_2$$

$$2Br^- + H_2O_2 + 2H^+ \rightarrow Br_2 + 2H_2O$$

Which substance is an intermediate?

A. Br_2
B. H_2O_2
C. H^+
D. O_2

383. Consider the following mechanism:

$$Br_2 + H_2O_2 \rightarrow 2Br^- + 2H^+ + O_2$$

$$2Br^- + H_2O_2 + 2H^+ \rightarrow Br_2 + 2H_2O$$

Which substance is a catalyst?

A. Br_2
B. H_2O_2
C. H^+
D. O_2

384. In careful calculations with calorimeters, the heat capacity of such items as the thermometer and the stirring rod must be considered. Suppose a coffee-cup calorimeter is used to calculate the heat of solution for sodium hydroxide, but the experimenter does *not* include the heat capacity of the thermometer and stirrer. How would this affect the results?

A. The calculated value of the heat of solution would be too low
B. The calculated value of the heat of solution would be too high
C. The calculated value of the heat of solution would be unaffected
D. The effect on the results depends on whether the reaction is endothermic or exothermic

385. An experimenter uses a coffee-cup calorimeter to measure the ΔH of a reaction. During the experiment, she leaves the lid off the cup, allowing a significant amount of convection. What effect does this have on the determination of the calculated magnitude of ΔH?

A. It is too low.
B. It is too high.
C. It is correct.
D. It may be too high or too low, depending on whether the reaction is endothermic or exothermic.

386. When heat is added to potassium chlorate, it generally disproportionates into potassium perchlorate and potassium chloride. If manganese dioxide is present during the process, however, the products are potassium chloride and oxygen gas. Which of the following is a *possible* explanation for these observations?

A. In the absence of manganese dioxide, the first set of products is more stable than the second. Manganese dioxide is a catalyst that stabilizes the second set of products.
B. Although the second set of products are more stable than the first, in the absence of manganese dioxide the second reaction occurs at a negligible rate. Manganese dioxide is a catalyst that increases the rate of the second reaction.
C. Manganese dioxide cannot be a catalyst, since it changes the products of the reaction; it must be actively involved with the second reaction.
D. Manganese dioxide cannot be a catalyst, since the presence of a catalyst would affect the reverse reaction as well; it must be a kinetic promoter of the second reaction.

387. For condensed phases, enthalpy change approximates internal energy change under which of the following conditions?

A. large volume changes
B. evaporation
C. sublimation
D. low to moderate pressure changes

388. Which of the following values can be measured with a coffee cup calorimeter?

I. internal energy
II. enthalpy
III. change in enthalpy

A. I only
B. II only
C. III only
D. I, II, and III

389. What does the subscript *f* indicate in ΔH_f^o?

I. The substance is formed from its constituent elements.
II. All reactants and products are at 1 bar (approx. 1 atm).
III. The reaction takes place at 25 °C.

A. I only
B. II only
C. I and II only
D. I, II, and III

390. What does the superscript 'o' indicate in ΔH_f^o?

I. The substance is formed from its constituent elements.
II. All reactants and products are at 1 bar (approx. 1 atm).
III. The reaction takes place at 25 °C.

A. I only
B. II only
C. I and II only
D. I, II, and III

Entropy

391. Which of the following 5 g samples of CO_2 has the highest entropy?

A. $CO_2(s)$
B. $CO_2(l)$
C. $CO_2(g)$
D. $CO_2(aq)$

392. Which of the following 5 g samples has the highest entropy?

A. $O(g)$
B. $O_2(g)$
C. $O_3(g)$
D. All of these have the same entropy

393. Which choice is ranked correctly from lowest to highest entropy per gram of NaCl?

A. $NaCl(g)$, $NaCl(aq)$, $NaCl(l)$, $NaCl(s)$
B. $NaCl(s)$, $NaCl(aq)$, $NaCl(l)$, $NaCl(g)$
C. $NaCl(s)$, $NaCl(l)$, $NaCl(aq)$, $NaCl(g)$
D. $NaCl(s)$, $NaCl(l)$, $NaCl(g)$, $NaCl(aq)$

394. Entropy:

I. is a state function.
II. is an extensive property.
III. has an absolute zero value.

A. I only
B. III only
C. I and II only
D. I, II, and III

395. Which of the following is/are valid statement(s) of the second law of thermodynamics?

I. Heat always flows from a warmer object to a cooler object
II. In a cyclic process, heat cannot be converted completely to work
III. The entropy of the universe never decreases

A. III only
B. I and II only
C. II and III only
D. I, II, and III

396. Is it possible for the entropy of a system to decrease when the system undergoes a spontaneous reaction?

A. Yes, but only if the entropy gain of the environment is greater in magnitude than the magnitude of the entropy loss in the system
B. Yes, but only if the reaction is endothermic
C. No, because this would violate the first law of thermodynamics
D. No, because this would violate the second law of thermodynamics

397. Is it possible for the entropy of the universe to decrease when the system undergoes a spontaneous reaction?

A. Yes, but only if the entropy gain of the environment is smaller in magnitude than the magnitude of the entropy loss in the system
B. Yes, but only if the reaction is exothermic
C. No, because this would violate the first law of thermodynamics
D. No, because this would violate the second law of thermodynamics

398. For the reaction:

$$2O(g) \rightarrow O_2(g)$$

what are the signs of the enthalpy and entropy changes?

A. Both are negative.
B. The enthalpy change is negative, while the entropy change is positive.
C. The enthalpy change is positive, while the entropy change is negative.
D. Both are positive.

Use the following information to answer questions 399-400. Two large heat reservoirs are connected by a thin metal bar. The first heat reservoir is 300 K. The second heat reservoir is 100 K.

399. As heat is transferred through the bar from the hot reservoir to the cooler one, which reservoir experiences the greatest change in entropy?

A. The cold reservoir because the same amount of energy change has a greater proportional effect.
B. The hot reservoir because it is losing energy.
C. There is no change in entropy because the system is isolated.
D. Since the same amount of energy leaving one enters the other, they both experience the same change in entropy.

400. What is the change in entropy when 300 joules of heat are conducted through the metal bar?

A. –2 J/K
B. 0 J/K
C. 2 J/K
D. 4 J/K

---

401. Which of the following is true concerning the entropy of the universe for a chemical reaction that has reached equilibrium?

A. The entropy is constantly changing.
B. The entropy has reached a maximum.
C. The entropy has reached a minimum.
D. The system has reached zero entropy.

Use the following information to answer questions 402-404. A man finds one hundred blocks scattered across the floor of an empty room. He neatly stacks the blocks in the center of the room.

402. Which of the following increases in entropy as a result of the man's actions?

I. the blocks
II. the man
III. the universe

A. II only
B. III only
C. II and III only
D. I, II, and III

403. Whenever entropy of the universe increases, some energy permanently loses some of its potential to do work. Which energy best represents this lost potential when the man stacks the blocks?

A. The potential energy of the blocks.
B. The heat energy created by the man.
C. The chemical energy of the nutrients in the man's body.
D. The kinetic energy of the air molecules in the room.

404. Put the following in order from greatest to smallest amount of energy:

I. The potential energy of the blocks.
II. The heat energy created by the man.
III. The chemical energy of the nutrients in the man's body.

A. I, II, III
B. II, III, I
C. III, I, II
D. III, II, I

---

405. Entropy increases with:

I. volume
II. number
III. temperature

A. II only.
B. III only
C. II and III only
D. I, II, and III

406. Consider the following equation for entropy change in a closed system:

$$\Delta S = \frac{q}{T}$$

Which of the following statements is false?

A. Entropy of the system decreases as heat is transferred out of the system.
B. Entropy changes due to heat transfer are greater at low temperatures.
C. The equation is only valid for a reversible process.
D. Entropy S increases as temperature T decreases.

407. The diagrams below represent glass tubes with black and white marbles. Which system has the greatest entropy?

A.

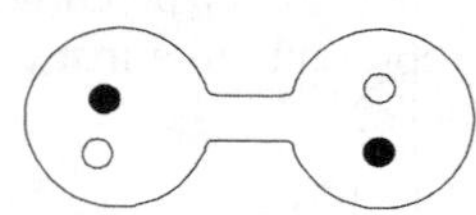

C.

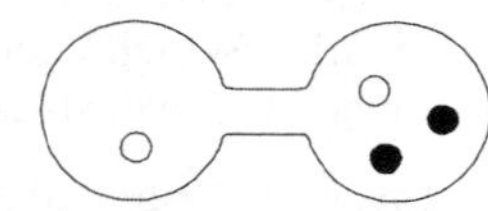

B.

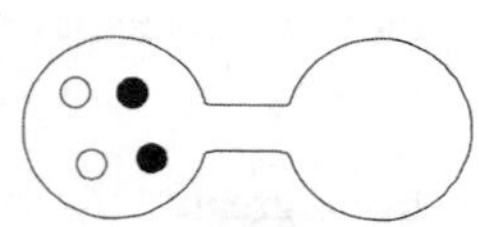

D.

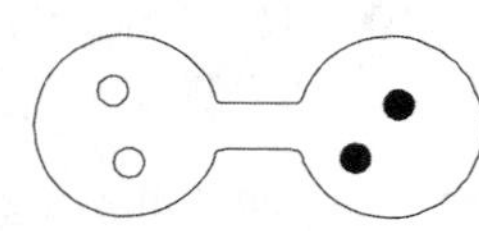

408. The standard molar entropy of O_2 gas is 0 J mol^{-1} K^{-1} at:

A. 0 K
B. 273 K
C. 298 K
D. The standard molar entropy of a substance doesn't depend upon temperature.

Use the table below to answer questions 409-410. The table below shows the standard molar entropy values for selected substances at 298 K.

SUBSTANCE	S^{o}, J/MOL-K
H_2	130.6
N_2	191.5
NH_3	192.5

409. What is the change in entropy for the production of one mole of ammonia from its constituent elements at 298 K?

A. −198.3 J/K
B. −99.2 J/K
C. 99.2 J/K
D. 198.3 J/K

410. The production of ammonia from its constituent elements is usually carried out at just under 800 K. At 800 K and 1 atm, the molar entropy of ammonia is:

A. less than 192.5 J/MOL-K
B. greater than 192.5 J/MOL-K
C. 192.5 J/MOL-K
D. May be either greater than or less than 192.5 J/MOL-K

Gibbs Free Energy

411. Which of the following conditions are required in order for a negative ΔG to indicate a spontaneous reaction?

I. Constant pressure
II. Constant temperature
III. Constant volume

A. I only
B. II only
C. I and II only
D. I, II, and III

412. Which of the following conditions are required for the formula: $\Delta G = \Delta H - T\Delta S$?

I. Constant pressure
II. Constant temperature
III. Constant volume

A. I only
B. II only
C. I and II only
D. I, II, and III

413. The Gibbs free energy change of a reaction is the maximum energy that the reaction will release to do:

A. any type of work.
B. non*P-V* work only.
C. *P-V* work only.
D. work and create heat.

414. Which of the following is/are true concerning Gibbs free energy?

I. The Gibbs free energy of the universe is conserved.
II. Gibbs free energy of a system is conserved.
III. Gibbs free energy does not obey the first law of thermodynamics.

A. I only
B. II only
C. III only
D. I and II only

415. A spontaneous reaction:

A. occurs quickly.
B. can do work on the surroundings.
C. releases heat.
D. cannot be endothermic.

416. In the formula: $\Delta G = \Delta H - T\Delta S$, which of the following is true concerning G, H, and S?

A. They refer to the system.
B. They refer to the surroundings.
C. They refer to the universe.
D. G refers to the universe, H to the surroundings, and S to the system.

Use the following information to answer questions 417-419. Under constant pressure conditions, the enthalpy change of a reaction represents the heat transferred into the system. If the enthalpy change is negative, heat is transferred to the surroundings. When heat is transferred to the surroundings, the entropy of the surroundings increases. The entropy change of the universe equals the negative of the Gibbs energy change of the system divided by the temperature. Putting this together gives:

$$\Delta G = \Delta H - T\Delta S$$

417. Which of the following statements is true of a reaction that obeys the formula $\Delta G = \Delta H - T\Delta S$?

A. If the reaction is exothermic, it must also be spontaneous.
B. The reaction will run to completion.
C. The reaction must take place under standard conditions.
D. If ΔG is negative, entropy of the universe will be increased as the system moves toward equilibrium.

418. Under which of the following conditions must a reaction be spontaneous?

A. Both enthalpy change and entropy change are negative.
B. Both enthalpy change and entropy change are positive.
C. Enthalpy change is negative and entropy change is positive.
D. Enthalpy change is positive and entropy change is negative.

419. If a reaction is exothermic but experiences a decrease in entropy of the system, under what conditions is the reaction most likely to be spontaneous?

A. low temperature
B. high temperature
C. low pressure
D. high pressure

420. Given the fact that the standard free energies of formation of water vapor and liquid water are –229 kJ/mol and –237 kJ/mol respectively, what is the free energy of the vaporization of water? Assume all measurements were preformed at 25°C.

A. –466 kJ/mol
B. –8 kJ/mol
C. +8 kJ/mol
D. +466 kJ/mol

421. Which of the following is required for a spontaneous reaction:

A. The entropy of the universe increases
B. The enthalpy of the universe increases
C. The free energy of the universe increases
D. The temperature of the universe increases

422. When the enthalpy of a system decreases, the *surroundings*:

A. Experience an increase in entropy and an increase in enthalpy
B. Experience an increase in entropy and a decrease in enthalpy
C. Experience a decrease in entropy and an increase in enthalpy
D. Experience a decrease in entropy and a decrease in enthalpy

423. A particular exothermic reaction is found to also produce a decrease in the entropy of the system. This reaction is:

A. Nonspontaneous at all temperatures
B. Spontaneous only at low temperatures
C. Spontaneous only at high temperatures
D. Spontaneous at all temperatures

424. A reaction is spontaneous at all temperatures. The reaction is:

A. exothermic and decreases the entropy of the system
B. exothermic and increases the entropy of the system
C. endothermic and decreases the entropy of the system
D. endothermic and increases the entropy of the system

425. A particular endothermic reaction causes a decrease in the entropy of the system. This reaction:

A. Can never occur, since it decreases system entropy
B. Can never occur, since the free energy change will be positive
C. Can only occur at low temperatures, where entropy is a smaller factor
D. Can occur if coupled to another reaction

426. The standard Gibbs free energy of formation for $H_2(g)$ at 25 °C is:

A. 0 kJ/mol
B. negative
C. positive
D. More information is needed to answer the question.

427. Consider the reaction:

$$A(g) + 2B(g) \rightarrow C(l)$$

In order for this reaction to be spontaneous under standard conditions, the reaction must:

A. result in an increase in entropy of the system.
B. take place in the presence of a catalyst.
C. be exothermic.
D. This reaction could not be spontaneous under standard conditions.

428. When 100 mL of a 2*M* NaCl solution and 100 mL of a 2*M* $AgNO_3$ solution are mixed, a precipitate is formed. Considering the two solutions to be the system:

A. The entropy of the system and the entropy of the surroundings both increase
B. The entropy of the system increases but the entropy of the surroundings decrease
C. The entropy of the system decreases but the entropy of the surroundings increases
D. The entropy of the system and the entropy of the surroundings both decrease

429. Suppose a scientist made the claim that all spontaneous reactions were exothermic. Which of the following would provide the strongest challenge to this claim?

A. An endothermic reaction that only proceeds when coupled to an exothermic reaction
B. An endothermic reaction that only proceeds at a reasonable rate when a catalyst is present
C. An endothermic reaction which is not spontaneous
D. An exothermic reaction which is not spontaneous

430. The standard free energies of formation (at 25°C) of the gas phase of several compounds are given below. If samples of the gases were placed in separate containers at 25°C and 1 atm and allowed to reach equilibrium, which container would end up with the greatest proportion of nonpolar gases?

Container	Gas	ΔG°
1	HF	174 kJ/mol
2	HCl	187 kJ/mol
3	NO	211 kJ/mol

A. 1
B. 2
C. 3
D. More information is needed

431. Under standard conditions, at what temperature would the following reaction be at equilibrium? ($\Delta H° = -484$ kJ/mol; $\Delta S° = -89$ J/mol K)

$$2H_2(g) + O_2(g) \rightarrow 2H_2O(g)$$

A. 0°C
B. 25°C
C. 100°C
D. 5200°C

Questions 432–437 refer to the following graph, which shows the temperature dependence of the standard free energy change for the reaction:

$$\tfrac{1}{2}O_2(g) + 2Ag(s) \rightarrow Ag_2O$$

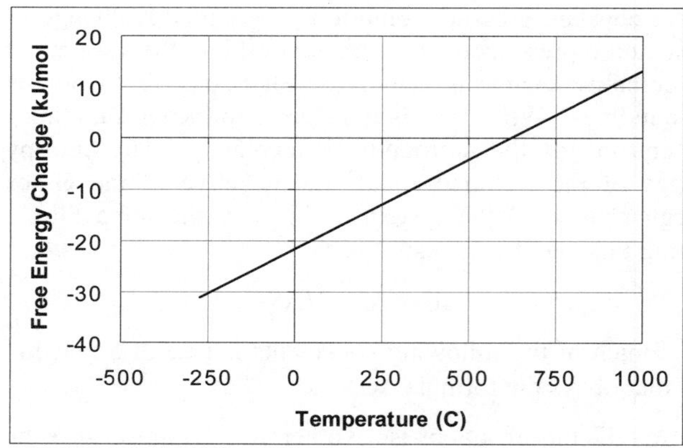

432. What is the standard free energy change of the following reaction at 0°C?

$$O_2(g) + 4Ag(s) \rightarrow 2Ag_2O$$

A. –44 kJ/mol
B. –11 kJ/mol
C. 0
D. +11 kJ/mol

433. What is the standard enthalpy change for the graphed reaction?

A. –31 kJ/mol
B. 0
C. +12 kJ/mol
D. Cannot be determined from the graph

434. What is the standard entropy change for the graphed reaction?

A. –35 J/(mol K)
B. 0
C. +12 J/(mol K)
D. +35 J/(mol K)

435. At what temperature does the graphed reaction have an equilibrium constant of 1?

A. −273°C
B. 0°C
C. 621°C
D. Cannot be determined from the information given

436. A researcher finds that, at a temperature of 750°C, solid silver and oxygen gas can be combined to form solid silver oxide. What can be concluded from this?

A. The results of the experimenter agree with the graph; the graph shows silver oxide production being spontaneous at standard conditions and 750°C
B. The experimenter has made an error; the graph clearly indicates that the reaction is non-spontaneous at this temperature
C. The experimenter probably performed the experiment under nonstandard conditions
D. The experiment probably produced less than one mole of silver oxide

437. An experimenter finds that, at room temperature, exposing solid silver to oxygen gas at a pressure of 1 atmosphere for 1 hour produced no observable reaction. What can be concluded from this?

A. The results of the experimenter agree with the graph, since the graph shows silver oxide production being nonspontaneous under standard conditions at room temperature
B. The graph is probably in error, since it shows silver oxide production being spontaneous under standard conditions at room temperature
C. The experimenter probably performed the experiment under nonstandard conditions
D. Although the reaction is spontaneous under standard conditions at room temperature, the rate is too low to produce a measurable effect in one hour

---

438. What is the standard Gibbs free energy of formation of water vapor at 25°C, if, for the reaction shown below under standard conditions, $\Delta H = -484$ kJ/mol and $\Delta S = -89$ J/mol K?

$$2H_2(g) + O_2(g) \rightarrow 2H_2O(g)$$

A. –457 kJ/mol
B. –395 kJ/mol
C. –229 kJ/mol
D. Water vapor does not form at 25 °C.

Solutions

439. Which of the following are possible?

A. a liquid solution resulting from a gaseous solute and a liquid solvent
B. a solid solution resulting from a solid solute and a solid solvent
C. a gaseous solution resulting from a gaseous solute and a gaseous solvent
D. A, B, and C are all possible.

440. Which of the following are solutions?

I. Sodium chloride in water
II. An alloy of 2% carbon and 98% iron
III. Water vapor in nitrogen gas

A. I only
B. I and II only
C. II and III only
D. I, II, and III

441. Which of the following are characteristics of an ideally dilute solution?

I. Solute molecules do not interact with each other.
II. Solvent molecules do not interact with each other.
III. The mole fraction of the solvent approaches 1.

A. I only
B. III only
C. I and III only
D. I, II, and III

442. Which of the following are characteristics of an ideal solution?

I. Solute molecules do not interact with each other.
II. Solvent molecules do not interact with each other.
III. Solvent-solute interactions are similar to solute-solute and solvent-solvent interactions.

A. I only
B. III only
C. I and II only
D. I, II, and III

443. Which compound is most likely to be more soluble in benzene (a nonpolar solvent) than in water?

A. Silver chloride
B. H_2S
C. SO_2
D. CO_2

444. Which of the following is the most soluble in benzene (C_6H_6)?

A. hydrobromic acid
B. octane (C_8H_{18})
C. sodium benzoate
D. sucrose ($C_6H_{12}O_6$)

445. Which of the following is most likely to be soluble in ammonia?

A. H_2
B. N_2
C. CO_2
D. SO_2

446. Why are most salts more soluble in water than in benzene?

A. Because the molecular mass of water is similar to the atomic mass of most ions
B. Because in order to dissolve a salt in benzene, the strong intermolecular attractions present in benzene must be disrupted
C. Because the dipole moment of water can help compensate for the loss of ionic bonding when the salt dissolves
D. Because benzene is aromatic, and thus very stable

447. Why is octane more soluble in benzene than in water?

A. Because octane and benzene have similar molecular weights
B. Because the bonds between octane and water are much weaker than the bonds between water molecules
C. Because the bonds between benzene and octane are much stronger than the bonds between water and octane
D. Because water cannot dissociate in the presence of octane

448. Hexane is infinitely soluble in octane. Why?

A. Because hexane and octane have similar molecular weights
B. Because the hexane-octane intermolecular bond is much stronger than either the hexane-hexane or the octane-octane intermolecular bonds
C. Because hexane can hydrogen bond to octane
D. Because the entropy increase upon mixing the two materials is the dominant factor in the free energy change when the two are mixed

449. 20 grams of an unknown substance are dissolved in 70 grams of water. When the solution is transferred to another container, it is found to weigh 92 grams. Which of the following is a possible explanation?

A. The reaction was exothermic, resulting in an increase in the average molecular speed
B. The solution reacted with the first container, causing some byproducts to be transferred with the solution
C. The solution reacted with the second container, causing a precipitate to form
D. Some of the solution was left in the first container

Colloids

450. Which of the following describes a colloidal suspension or colloid?

A. Tiny clay particles dispersed in water that settle out over time
B. Hemoglobin molecules in the cytosol
C. Carbon dioxide molecules mixed in a soft drink
D. A saturated solution of silver chloride

451. All of the following are methods which may extract colloidal particles except:

A. adding an electrolyte
B. dialysis
C. heating
D. simple filtration

452. A beam of light is shone through a colloid and a solution. Which of the following is true concerning the Tyndall effect and colloids?

A. The path of the beam is *visible* in both the colloid and the solution.
B. The path of the beam is *invisible* in both the colloid and the solution.
C. The path of the beam is *visible* in the colloid but *invisible* in the solution.
D. The path of the beam is *invisible* in the colloid but *visible* in the solution.

453. What causes colloidal particles to settle out during coagulation?

A. Particles bind together and settle due to gravity.
B. Particles break apart and settle due to gravity.
C. Particles bind to the walls of a container.
D. Particles emulsify and settle out.

Electrolytes

454. The chlorite ion is which of the following?

A. ClO^-
B. ClO_2^-
C. ClO_3^-
D. ClO_4^-

455. The bicarbonate ion is which of the following?

A. CO_3^-
B. CO_3^{2-}
C. HCO_3^-
D. HCO_3^{2-}

456. Which answer choice gives the following in the correct order: nitrate, nitrite, sulfate, sulfite?

A. NO_2^-, NO_3^-, SO_3^-, SO_4^-
B. NO_3^-, NO_2^-, SO_3^-, SO_4^-
C. NO_3^-, NO_2^-, SO_4^-, SO_3^-
D. NO_2^-, NO_3^-, SO_3^-, SO_4^-

457. Hydration involves:

A. the breaking of water-water bonds.
B. the breaking of solute-solute bonds.
C. the formation of solute-water bonds.
D. both the breaking of water-water bonds and the formation of solute-water bonds.

458. The hydration number of an ion is:

A. the number of ions that can dissolve in one liter of water.
B. the number of ions that bond to a water molecule when in aqueous solution.
C. the number of water molecules required to dissolve one liter of ions.
D. the number of water molecules that bond to an ion in an aqueous solution.

459. Which of the following is a good test for an electrolyte?

A. Place two leads into the substance, and connect the leads with a resistor. If the substance is an electrolyte, a current will flow.
B. Place two leads into the substance, and connect the leads with a battery. If the substance is an electrolyte, a current will flow.
C. Dissolve the substance in water. Place two leads into the solution, and connect the leads with a resistor. If the substance is an electrolyte, a current will flow.
D. Dissolve the substance in water. Place two leads into the solution, and connect the leads with a battery. If the substance is an electrolyte, a current will flow.

460. Which of the following is/are strong electrolytes?

I. Salts
II. Strong Bases
III. Weak Acids

A. I only
B. I and II only
C. I and III only
D. I, II, and III

461. Potassium oxide contains:

A. covalent bonds and is a non-electrolyte.
B. covalent bonds and is a strong electrolyte.
C. ionic bonds and is a weak electrolyte.
D. ionic bonds and is a strong electrolyte.

462. Hydrochloric acid contains:

A. covalent bonds and is a non-electrolyte
B. covalent bonds and is a strong electrolyte
C. ionic bonds and is a weak electrolyte
D. ionic bonds and is a strong electrolyte

463. Which of the following is a non-electrolyte?

A. Sucrose
B. Ammonia
C. Sodium chloride
D. Potassium hydroxide

464. Which of the following statements is correct?

A. Strong electrolytes are highly soluble in water.
B. Weak electrolytes are insoluble in water.
C. The extent to which an electrolyte dissolves in solution does not indicate whether it is strong or weak.
D. Most electrolytes are insoluble in water.

Use the chart below to answer questions 465-468. The flowchart below allows for classification of electrolytes.

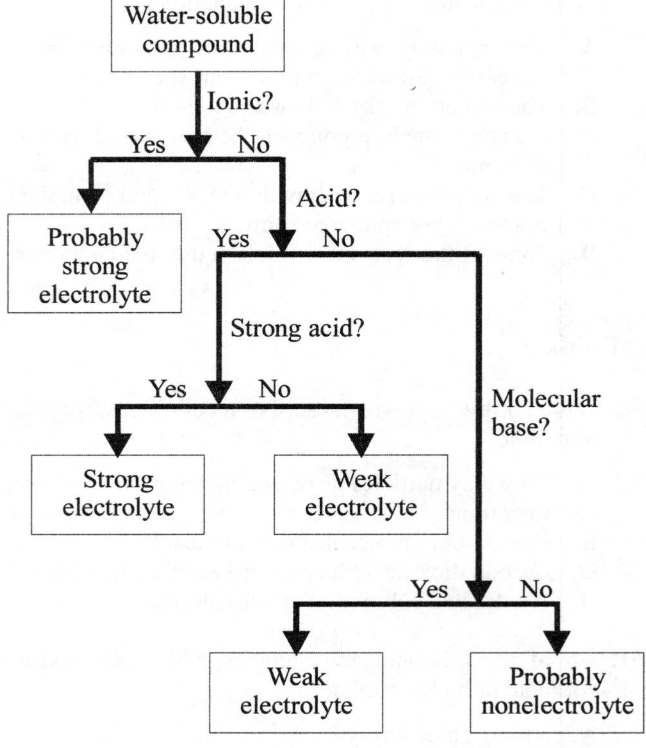

465. According to the flowchart, NH_3 is:

A. a strong electrolyte
B. a weak electrolyte
C. probably not an electrolyte
D. an acid

466. Which of the following is true according to the flowchart?

A. All ionic compounds are strong electrolytes.
B. All acids are strong electrolytes.
C. Molecular bases are weak electrolytes.
D. Non molecular bases are probably not electrolytes.

467. Compound X is a weak electrolyte. $Mg(OH)_2$ is more soluble in water when compound X is added. Compound X is most likely:

A. ionic
B. a molecular base
C. a weak acid
D. a strong acid

468. The flow chart indicates that strong acids are strong electrolytes and weak acids are weak electrolytes. Why?

- **A.** Strong acids are more soluble than weak acids.
- **B.** The ions of weak acids carry smaller charge in solution.
- **C.** Weak acids don't produce as many ions.
- **D.** Strong acids are more reactive.

Solubility Guidelines

469. Ionic compounds containing alkali metals are soluble in water. Other than those, which of the following types of compounds are insoluble in water?

- **A.** compounds containing the nitrate ion
- **B.** sulfate compounds other than those containing Ca^{2+}, Sr^{2+}, Ba^{2+}, Hg_2^{2+}, and Pb^{2+}
- **C.** compounds containing the carbonate ions excluding ammonium salts.
- **D.** compounds containing the ammonium ion

470. Which of the following metal ions forms a soluble compound with Cl^-, Br^-, and I^-?

- **A.** Al^{3+}
- **B.** Ag^{2+}
- **C.** Hg_2^{2+}
- **D.** Pb^{2+}

471. Which of the following metal ions forms an insoluble compound with OH^-?

- **A.** Ca^{2+}
- **B.** Al^{3+}
- **C.** Sr^{2+}
- **D.** Ba^{2+}

472. Which of the following metal ions forms an insoluble compound with S^{2-}?

- **A.** NH^{4+}
- **B.** Al^{2+}
- **C.** Sr^{2+}
- **D.** Ba^{2+}

Units of Concentration

473. Which of the following is/are proper procedure(s) to follow when making a 2.000 *M* solution of sodium chloride?

- **I.** Measure out 116.88 grams of sodium chloride onto a tared weighing paper, transfer all of the sodium chloride into a 1.0000 L volumetric flask, and then add water up to the mark.
- **II.** Measure out 29.22 grams of sodium chloride onto a tared weighing paper, transfer all of the sodium chloride into a 250.0 mL volumetric flask, and then add water up to the mark.
- **III.** Fill a 500.0 mL volumetric flask to the mark, weigh 58.44 grams of sodium chloride onto a tared weighing paper, and then transfer the sodium chloride into the flask.

- **A.** I only
- **B.** III only
- **C.** I and II only
- **D.** I, II, and III

474. Which of the following is a proper procedure for making a 2.000 *m* solution of sodium chloride?

- **I.** Measure out 116.88 grams of sodium chloride onto a tared weighing paper, transfer all of the sodium chloride into a 1.0000 L volumetric flask, and then add water up to the mark.
- **II.** Measure out 29.22 grams of sodium chloride onto a tared weighing paper, transfer all of the sodium chloride into a 250.0 mL volumetric flask, and then add water up to the mark.
- **III.** Fill a 500.0 mL volumetric flask to the mark, weigh 58.44 grams of sodium chloride onto a tared weighing paper, and then transfer the sodium chloride into the flask.

- **A.** I only
- **B.** III only
- **C.** I and II only
- **D.** I, II, and III

475. Which of the following measures of concentration is/are NOT affected by changes in temperature?

- **I.** Molarity
- **II.** Molality
- **III.** Mole Fraction

- **A.** I only
- **B.** III only
- **C.** II and III
- **D.** I, II, and III

476. 18.0 grams of sucrose (molecular weight 180) are dissolved in 1.8 kg of water at room temperature. What is the molality of the sucrose solution? (Assume the sucrose does not significantly change the volume of the solution.)

A. 0.056 *m*
B. 0.091 *m*
C. 0.100 *m*
D. 0.180 *m*

477. 18.0 grams of sucrose (molecular weight 180) are dissolved in 1.8 kg of water at room temperature. What is the molarity of the sucrose solution? (Assume the sucrose does not significantly change the volume of the solution.)

A. 0.056 *M*
B. 0.091 *M*
C. 0.100 *M*
D. 0.180 *M*

478. 18.0 grams of sucrose (molecular weight 180) are dissolved in 1.8 kg of water at room temperature. What is the mole fraction of sucrose in the sucrose solution? (Assume the sucrose does not significantly change the volume of the solution.)

A. 0.0010
B. 0.0056
C. 0.0091
D. 0.0180

479. 18.0 grams of sucrose (molecular weight 180) are dissolved in 1.44 kg of room temperature ethanol (molecular weight 46). What is the molarity of the sucrose solution? (Assume the density of the resulting solution to be 0.80 g/cm^3.)

A. 0.056 *M*
B. 0.091 *M*
C. 0.100 *M*
D. 0.180 *M*

480. What is the molarity of pure water?

A. 1.0
B. 7.0
C. 18.0
D. 55.6

481. What is the molality of pure water?

A. 1.0
B. 7.0
C. 18.0
D. 55.6

482. What is the mole fraction of water in pure water?

A. 1.0
B. 7.0
C. 18.0
D. 55.6

483. What is the concentration of the chloride ion when enough water is added to 12 grams of sodium chloride to make 4.0 L of solution?

A. 0.025 *M*
B. 0.05 *M*
C. 0.1 *M*
D. 0.2 *M*

484. What is the concentration of chloride ion in a solution in which enough water was added to 11 grams of calcium chloride to make 4.0 L of solution?

A. 0.025 *M*
B. 0.05 *M*
C. 0.1 *M*
D. 0.2 *M*

485. What is the concentration of bromide ion in a solution in which 35 mL of 0.40 *M* NaBr is mixed with 50mL of 0.10 *M* $CaBr_2$?

A. 0.10 *M*
B. 0.20 *M*
C. 0.28 *M*
D. 0.40 *M*

486. 30 grams of NaOH are dissolved in 100 grams of water. What is the molality of NaOH in the resulting solution?

A. 0.0075 *m*
B. 0.30 *m*
C. 7.5 *m*
D. 300 *m*

487. What is the concentration of lithium ions in a 0.003 *M* solution of lithium carbonate?

A. 0.0015 *M*
B. 0.003 *M*
C. 0.006 *M*
D. 0.009 *M*

488. Which of the following describes mass percent?

A. 100 times grams of solute divided by moles of solvent
B. 100 times grams of solute divided by grams of solvent
C. 100 times grams of solute divided by moles of solution
D. 100 times grams of solute divided by grams of solution

489. Which of the following describes parts per million (ppm)?

- **A.** 1,000,000 times grams of solute divided by moles of solvent
- **B.** 1,000,000 times grams of solute divided by grams of solvent
- **C.** 1,000,000 times grams of solute divided by moles of solution
- **D.** 1,000,000 times grams of solute divided by grams of solution

490. In aqueous solutions, parts per million (ppm) is equal to:

- **A.** grams of solute per liter of solution.
- **B.** milligrams of solute per liter of solution.
- **C.** moles of solute per liter of solution.
- **D.** millimoles of solute per liter of solution.

491. If a container of water contains 0.06 mg of arsenic per liter, what is the parts per billion (ppb) of arsenic in water?

- **A.** 0.06 ppb
- **B.** 6 ppb
- **C.** 60 ppb
- **D.** 600 ppb

492. 0.1 mole of NaCl is placed in a fish tank containing 60 liters of water. What is the ppm of NaCl in the water in the fish tank?

- **A.** 0.00097 ppm
- **B.** 0.097 ppm
- **C.** 97 ppm
- **D.** 970 ppm

Solution Formation

493. The heat of hydration for any solute is:

- **A.** always positive.
- **B.** always negative.
- **C.** always zero.
- **D.** may be either positive or negative depending upon the solute.

494. When salt X is dissolved in water to form a 1 molar unsaturated solution, the temperature of the solution is seen to drop. From this information, we can conclude that for dissolving this salt in water:

- **A.** $\Delta H°$, $\Delta S°$, and $\Delta G°$ are all positive
- **B.** $\Delta H°$, $\Delta S°$, and $\Delta G°$ are all negative
- **C.** $\Delta H°$ is positive; $\Delta G°$ is negative
- **D.** $\Delta H°$ is positive; no conclusion can be reached about $\Delta G°$

495. The heat of a solution measures the energy absorbed in the:

- **A.** breaking of solvent-solvent bonds only.
- **B.** breaking of solute-solute bonds only.
- **C.** formation of solvent-solute bonds only.
- **D.** All of the above.

496. When solid potassium chloride is dissolved in water, the energies of the bonds formed are lower than the energies of the bonds broken. Why does the reaction proceed anyway?

- **A.** Undissolved potassium chloride compensates for the portion that dissolves
- **B.** The reaction does not take place under standard conditions
- **C.** The electronegativity of the water is increased by the interaction with potassium and chloride ions
- **D.** The increased disorder due to mixing results in an increased entropy of the system

497. What does a negative heat of solution indicate about solute-solvent bonds as compared to solute-solute bonds and solvent-solvent bonds?

- **A.** Solute-solute and solvent-solvent bonds are *stronger* than solute-solvent bonds.
- **B.** Solute-solute and solvent-solvent bonds are *weaker* than solute-solvent bonds.
- **C.** Solute-solute and solvent-solvent bond strength is equal to solute-solvent bond strength.
- **D.** No comparison of bond strength can be made based upon heat of solution.

498. Which of the following types of solutions is likely to exhibit no enthalpy change upon formation?

- **A.** An ideally dilute solution
- **B.** An ideal solution
- **C.** A real solution
- **D.** A solution held at constant temperature during formation

Use the information below to answer questions 499-503. The heat of hydration for NaCl is –783 kJ/mol. The change in enthalpy for the formation of solid NaCl from its ions in gaseous state is –786 kJ/mol.

499. What is the heat of solution when NaCl is dissolved in water?

- **A.** –3 kJ/mol
- **B.** +3 kJ/mol
- **C.** –1569 kJ/mol
- **D.** +1586 kJ/mol

500. Is the dissolution of NaCl in water a spontaneous process?

A. Yes, because the change in enthalpy is positive.
B. No, because it is an endothermic reaction.
C. It depends upon the temperature because the change in entropy is positive.
D. It depends upon the temperature because the change in enthalpy depends upon temperature.

501. Based on the heat of hydration of NaCl, what can be said about the hydrogen bonds of water compared to the bonds between the water molecules and the ions of NaCl?

A. The hydrogen bonds are stronger because hydrogen bonds are the strongest dipole-dipole forces.
B. The hydrogen bonds are stronger because the heat of hydration is negative.
C. The water-ion bonds are stronger because they are ionic bonds.
D. The water-ion bonds are stronger because the heat of hydration is negative.

502. Which of the following will increase the concentration of NaCl that can be dissolved in water?

A. increasing the temperature
B. decreasing the temperature
C. adding more NaCl to solution
D. adding more water to solution

503. Which energy diagram best represents the dissolution of NaCl in water?

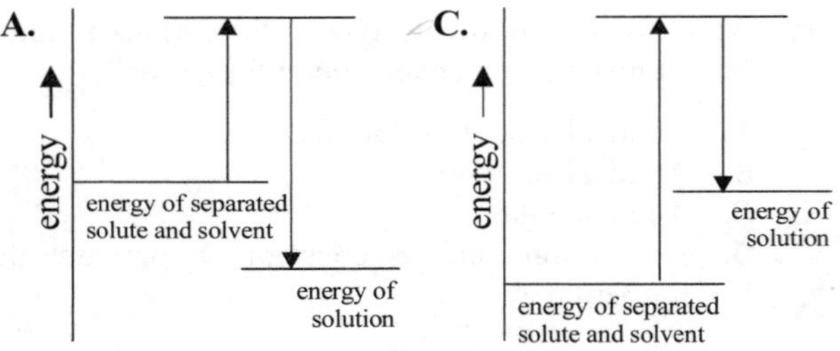

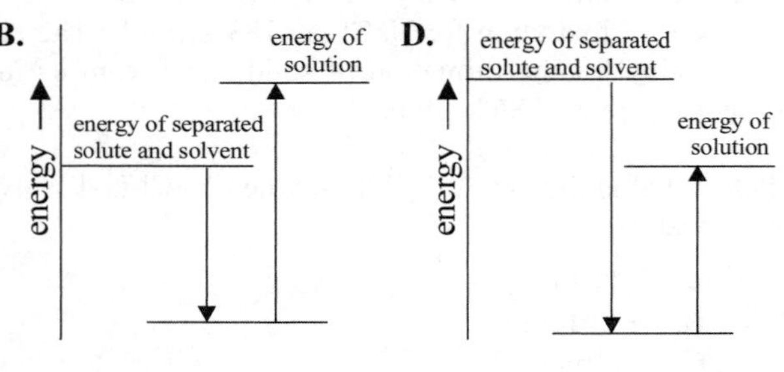

504. Which of the following explains why oil doesn't dissolve in water?

A. The entropy increase of the water as it solvates the oil is much greater than the entropy decrease of the oil upon salvation.
B. The entropy decrease of the water as it solvates the oil is much greater than the entropy increase of the oil upon salvation.
C. The energy *released* by the separation of water molecules is much *less* than the energy *absorbed* by the bonding of oil molecules with water.
D. The energy *absorbed* by the separation of water molecules is much *less* than the energy *released* by the bonding of oil molecules with water.

Vapor Pressure

505. Which of the following is a viable experimental method for determining the vapor pressure of a pure liquid?

A. Determine the greatest weight per area the surface of the liquid will support.
B. Measure the difference in height between the two sides of a manometer filled with the liquid.
C. Measure the rate at which the liquid evaporates.
D. Allow the liquid to reach equilibrium with its vapor and measure the partial pressure of the vapor.

506. Under what conditions does a solid sublimate?

A. The vapor pressure of the solid must be greater than the vapor pressure of the liquid at a particular temperature
B. The vapor pressure of the solid must be greater than the partial pressure of its vapor above it
C. The vapor pressure of the solid must be greater than the total pressure of the gas above it
D. The vapor pressure of the liquid must be greater than the total pressure of the gas above it

507. Under what conditions does a liquid boil?

A. The vapor pressure of the liquid must be greater than 100 torr.
B. The vapor pressure of the liquid must be greater than the partial pressure of its vapor above it.
C. The vapor pressure of the liquid must be greater than the total pressure of the gas above it.
D. The partial pressure of the vapor must be greater than the total pressure of the gas.

508. A vapor pressure curve for a volatile liquid is presented below. What is the approximate normal boiling point of the liquid?

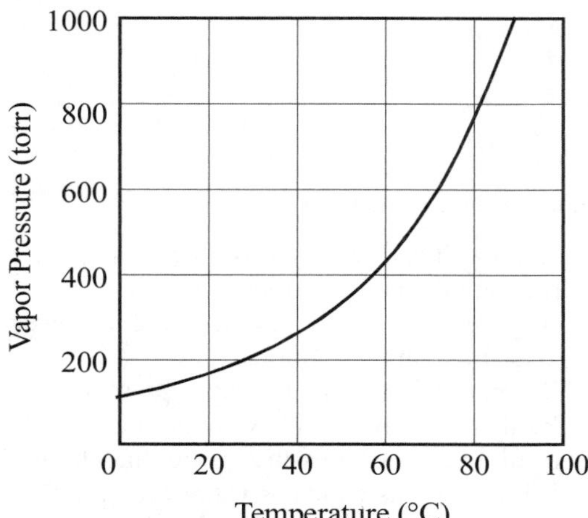

A. 23°C
B. 79°C
C. 100°C
D. 212°C

509. A student has a sample of bromine and a sample of iodine, contained in identical flasks. Bromine has a much higher vapor pressure than iodine. In order to make the vapor pressures of the two substances equal, the student could:

A. remove some of the bromine.
B. let the bromine expand into a larger volume.
C. heat the iodine.
D. there is no way to make the vapor pressures of the two substances equal.

510. On a certain day, water condenses on a surface maintained at 13°C, and boils when heated to 97°C. What can be concluded about the atmospheric conditions on that day?

A. The partial pressure of water in the atmosphere is greater than the vapor pressure of water at 13°C, and the total atmospheric pressure is greater than 760 torr.
B. The partial pressure of water in the atmosphere is less than the vapor pressure of water at 13°C, and the total atmospheric pressure is greater than 760 torr.
C. The partial pressure of water in the atmosphere is at least as great as the vapor pressure of water at 13°C, and the total atmospheric pressure is less than 760 torr.
D. The partial pressure of water in the atmosphere is at least as low as the vapor pressure of water at 13°C, and the total atmospheric pressure is less than 760 torr.

511. All of the following are true concerning vapor pressure EXCEPT:

A. For the same substance at a given pressure, the solid has a lower vapor pressure than the liquid.
B. For the same substance below the melting point, the solid has a higher vapor pressure than the liquid.
C. The normal melting point is the temperature at which the solid and liquid have equal vapor pressures at one atmosphere.
D. At 1 atm, a liquid is only at equilibrium with its vapor pressure at 100 °C.

512. If a liquid can achieve equilibrium with its vapor, why do puddles evaporate?

A. Because liquid molecules are carried away by air currents shifting the equilibrium.
B. Because vapor molecules are carried away by air currents shifting the equilibrium.
C. Because the sun turns the liquid in the puddles into steam.
D. Because the sun turns the vapor above the puddles into steam.

513. What is the vapor pressure of a 2 *m* sodium chloride solution at 100°C?

A. 100 torr
B. 709 torr
C. 760 torr
D. 787 torr

514. Which of the following solutions has the greatest vapor pressure?

A. 20 mL of pure water at 30°C
B. 30 mL of pure water at 20°C
C. 40 mL of 3 *M* aqueous sodium chloride at 25°C
D. 50 mL of 3 *M* aqueous glucose at 25°C

515. At 25°C, water has a vapor pressure of 23.8 torr and ethanol has a vapor pressure of 58.9 torr. If 50 grams of ethanol (M.W. 46) is added to 200 grams of water, what is the approximate vapor pressure of the mixture?

A. 5.3 torr
B. 21.7 torr
C. 27.0 torr
D. 82.7 torr

516. Which of the following forms of water has the highest vapor pressure?

A. ice
B. liquid water
C. water vapor under 100 °C
D. steam

517. The graph below shows the vapor pressures of the liquid and solid phase of the same substance.

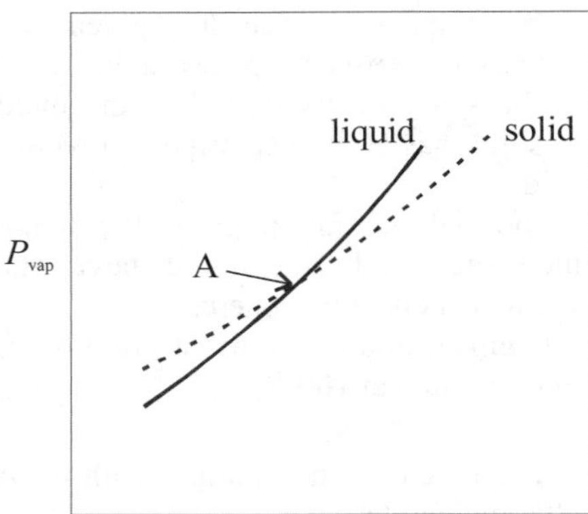

Which of the following is false?

A. The point A indicates the melting point of the substance.
B. The vapor pressure of the solid changes more slowly with temperature.
C. The results on the graph cannot be obtained at constant atmospheric pressure for normal phases.
D. At point A, the vapor pressure equals the atmospheric pressure.

518. Which of the following is true concerning the hard boiling of an egg?

A. The egg will become hardboiled faster at low altitudes.
B. The water will boil sooner at low altitudes.
C. The water will get hotter at high altitudes.
D. The water will reach a higher vapor pressure at high altitudes.

519. A solution of ethanol and water is slightly nonideal; the water-ethanol attractions, in particular, are stronger than the ethanol-ethanol attractions. This results in a lowering of the vapor pressure of the alcohol in the mixture beyond what is predicted by Raoult's law. Suppose a scientist measures the vapor pressure of ethanol above an ethanol-water mixture, and then uses Raoult's law to estimate the mole fraction of ethanol in the mixture. How is the estimate affected by the deviation from ideality?

A. The estimate is too low.
B. The estimate is too high.
C. The estimate is correct; the vapor pressure of the pure and mixed solutions are both depressed by the same percentage, leaving the mole fraction unchanged.
D. The estimate is too low if the mole fraction of ethanol is below 50%, and too high otherwise.

520. 58 grams of NaCl are added to 1 liter of pure water. The solution is heated to 100 °C. What is the vapor pressure?

A. 380 torr
B. 747 torr
C. 760 torr
D. 773 torr

521. Which of the following is true?

A. A pure liquid will always have a greater vapor pressure than its solutions.
B. A pure liquid will always have a lower vapor pressure than its solutions.
C. If liquid X has a greater vapor pressure than liquid Y, then adding liquid X to liquid Y will always result in a vapor pressure greater than liquid Y.
D. If liquid X has a greater vapor pressure than liquid Y, adding liquid X to liquid Y may result in a vapor pressure greater or lower than liquid Y.

522. Glass X contains pure water and glass Y contains an aqueous salt solution. The glasses are placed in a sealed container as shown below.

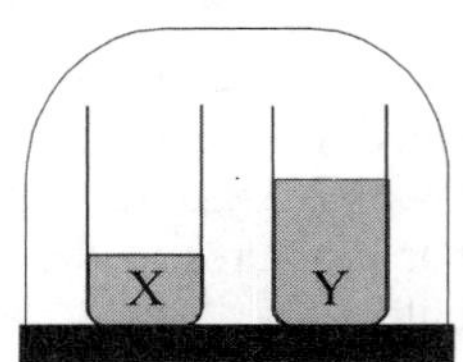

Which of the following represents the system after it has come to equilibrium?

A.

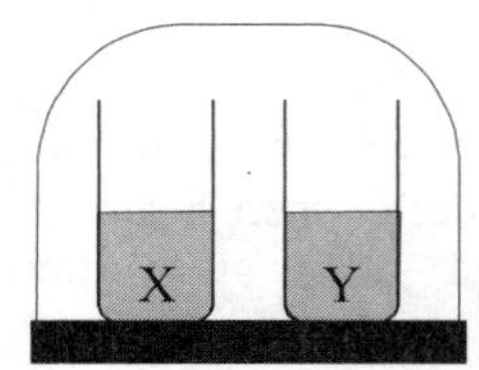

C.

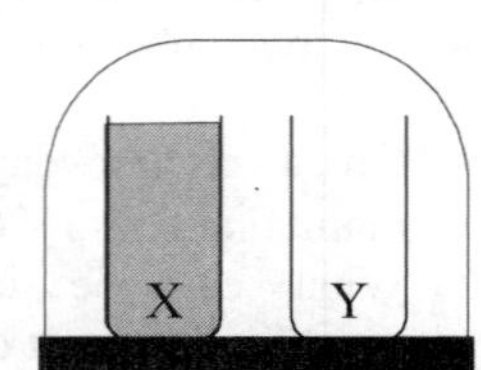

B.

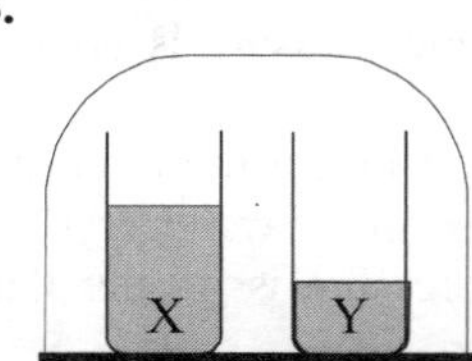

D.

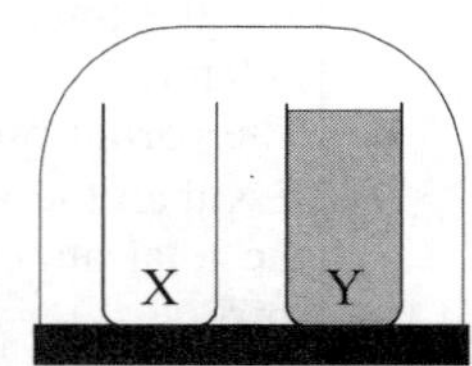

523. A nonvolatile solute is a solute that:

A. will not burn.
B. has no vapor pressure.
C. is nonreactive.
D. does not dissolve in water.

524. As atmospheric pressure increases, vapor pressure:

A. increases.
B. decreases.
C. is not affected.
D. The change in the vapor pressure depends upon the solution.

525. A pure volatile liquid is placed in a sealed container. Which of the following accurately describes how an equilibrium is reached between the rate of evaporation and rate of condensation?

A. The rate of condensation rises and the rate of evaporation falls until the rates are equal.
B. The rate of condensation falls and the rate of evaporation rises until the rates are equal.
C. The rate of condensation rises and the rate of evaporation remains constant until the rates are equal.
D. The rate of condensation remains constant and the rate of evaporation falls until the rates are equal.

Use the diagram below to answer questions 526-528. The diagram below shows a barometer filled with an unknown fluid at 25 °C. The barometer is made using the following steps: a tube is filled completely with the unknown fluid. The open end of the tube is sealed, and the tube is inverted. The open end is submerged in the unknown fluid and unsealed.

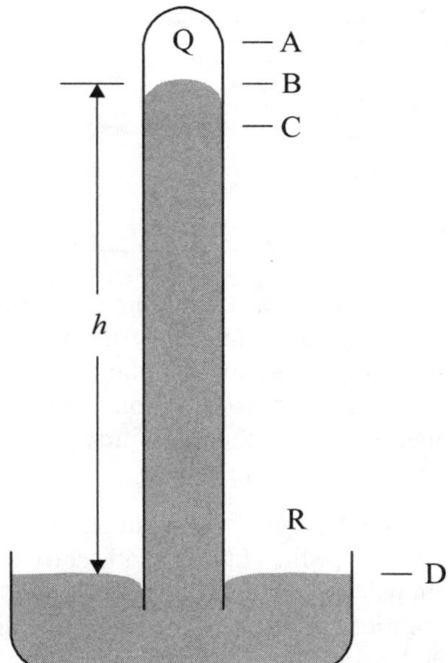

526. If the air surrounding the barometer is at 1 atm, which of the following describes the pressure at Q?

A. The pressure at Q is due only to the vapor pressure of the unknown fluid.
B. The pressure at Q is due to the vapor pressure of the unknown fluid and to the force of the liquid pushing upward.
C. The pressure at Q is 1 atm due to the surround air.
D. The pressure at Q is due to the vapor pressure plus 1 atm due to the surrounding air.

527. The unknown fluid has a vapor pressure of 12 torr and the same density as water. At 25 °C, water has a vapor pressure of 24 torr. Suppose that the unknown fluid were replaced with water. The fluid level would:

A. rise toward A.
B. remain at B.
C. fall toward C.
D. fall to D.

528. If the air surrounding the barometer were at 1 atm, and the fluid level were at level D, the vapor pressure of the unknown fluid would be:

A. less than 1 atm
B. 1 atm
C. greater than 1 atm
D. The fluid level could not be at level D.

Solubility

529. Which of the following is least likely to be soluble in water?

A. K_2CO_3
B. NH_4NO_3
C. MgO
D. RbI

530. An ion in solution that does not take part in the reaction is called a:

A. common ion.
B. spectator ion.
C. cation
D. ligand

531. Which of the following is NOT true concerning the solubility product, K_{sp}?

A. Its value depends upon temperature.
B. Its value depends upon the solvent.
C. Its value is independent of the concentration of the reactants.
D. Its value is zero for insoluble compounds.

Use the solubility table below to answer questions 532-540.

Solid	K_{sp}	Solid	K_{sp}
$NiCO_3$	$1.4x10^{-7}$	$Ba(OH)_2$	$5.0x10^{-3}$
$CaCO_3$	$8.7x10^{-9}$	$Ca(OH)_2$	$1.3x10^{-6}$
$BaCO_3$	$1.6x10^{-9}$	$Fe(OH)_2$	$1.8x10^{-15}$
$CuCO_3$	$2.5x10^{-10}$	$Ni(OH)_2$	$1.6x10^{-16}$
$MnCO_3$	$8.8x10^{-11}$	$BaCrO_4$	$8.5x10^{-11}$
$FeCO_3$	$2.1x10^{-11}$	FeS	$3.7x10^{-19}$
Ag_2CO_3	$8.1x10^{-12}$	NiS	$3.0x10^{-21}$

532. A certain textbook defines insoluble compounds as those compounds with a solubility less than 0.01 mol/L. Which of the following would be considered soluble by this definition?

I. $NiCO_3$
II. $Ba(OH)_2$
III. FeS

A. I only
B. II only
C. II and III only
D. I, II, and III

533. If we consider compounds with a solubility of less than 0.01 mol/L to be insoluble, and we use the K_{sp} for $Ca(OH)_2$ as given in the table and make standard calculations, we find that $Ca(OH)_2$ should be insoluble. However, $Ca(OH)_2$ is, in fact, soluble by the definition given. How can this discrepancy be reconciled?

A. Ca^{2+} ions bond with OH^- ions to form ion pairs in solution.
B. The reaction is exothermic, raising the temperature of the solution and increasing solubility.
C. Calcium hydroxide dissociates into three particles, not two.
D. Calcium hydroxide creates a basic solution, shifting the equilibrium expression to the right.

534. Comparing the K_{sp} values in the table, which of the following is the *least* soluble in water?

A. $MnCO_3$
B. $FeCO_3$
C. Ag_2CO_3
D. $BaCrO_4$

535. Comparing the K_{sp}s, which of the following is the *most* soluble in 0.1 *M* silver nitrate solution?

A. $MnCO_3$
B. $FeCO_3$
C. Ag_2CO_3
D. $BaCrO_4$

536. The solubility of $Ca(OH)_2$ in water:

A. is greatest in acidic solutions.
B. is greatest in neutral solutions.
C. is greatest in basic solutions.
D. does not change with pH.

537. The solubility product of $Ca(OH)_2$ in water:

A. is greatest in acidic solutions.
B. is greatest in neutral solutions.
C. is greatest in basic solutions.
D. does not change with pH.

538. NiI_2 is dissolved into a solution containing equal concentrations of OH^-, CO_3^{2-}, and S^{2-}. According to the information in the table, in what order will precipitation occur?

A. OH^-, CO_3^{2-}, then S^{2-}
B. OH^-, S^{2-}, then CO_3^{2-}
C. S^{2-}, OH^-, then CO_3^{2-}
D. S^{2-}, CO_3^{2-}, then OH^-

539. What is the solubility of $BaCO_3$ in water?

A. $1.6x10^{-9}$ mol/L
B. $1.4x10^{-6}$ mol/L
C. $4.0x10^{-5}$ mol/L
D. $8.1x10^{-3}$ mol/L

540. What is the solubility of $Ca(OH)_2$ in a 0.1 *M* solution of KOH?

A. $1.1x10^{-2}$ mol/L
B. $1.3x10^{-4}$ mol/L
C. $1.3x10^{-6}$ mol/L
D. $2.4x10^{-8}$ mol/L

---

541. Calculations of ion concentrations using the solubility product often deviate from experimental results. This is due to ion pairing in solution and to hydrolysis reaction undergone by some anions and cations. How are calculations using the solubility product likely to differ from experimental results?

A. Solubilities found by experiment are likely to be *lower* because the equilibrium is shifted to the *right* by ion pairing and hydrolysis reactions.
B. Solubilities found by experiment are likely to be *lower* because the equilibrium is shifted to the *left* by ion pairing and hydrolysis reactions.
C. Solubilities found by experiment are likely to be *higher* because the equilibrium is shifted to the *right* by ion pairing and hydrolysis reactions.
D. Solubilities found by experiment are likely to be *higher* because the equilibrium is shifted to the *left* by ion pairing and hydrolysis reactions.

542. When 0.20 *M* sodium chloride and 0.30 *M* silver nitrate are mixed, a white substance appears which can be isolated through centrifugation. The substance is most likely:

A. $NaNO_3$
B. AgCl
C. AgNa
D. NaCl

543. Which of the following is the correct equilibrium expression for the dissolution of calcium phosphate in water?

A. $K_{sp} = [Ca^{2+}][PO_4^{3-}]$
B. $K_{sp} = [Ca^{2+}]^2[PO_4^{3-}]^3$
C. $K_{sp} = [Ca^{2+}]^3[PO_4^{3-}]^2$
D. $K_{sp} = [3Ca^{2+}][2PO_4^{3-}]$

Use the information below to answer questions 544-547. The solubility of CaF_2 is increased by adding acid to the solution. The process can be understood in terms of the following summation of reactions:

$$CaF_2(s) \rightleftharpoons Ca^{2+}(aq) + 2F^-(aq)$$
$$2\{F^-(aq) + H^+(aq) \rightleftharpoons HF(aq)\}$$
$$CaF_2(s) + 2H^+(aq) \rightleftharpoons Ca^{2+}(aq) + 2HF\ (aq)$$

544. A saturated solution of CaF_2 is created by adding excess CaF_2 to water. The excess CaF_2 precipitate is filtered out of solution. Is the solution still saturated?

A. Yes, because if any CaF_2 is added, precipitate will form.
B. Yes, because the concentration of the solution hasn't changed.
C. No, because a saturated solution must be at equilibrium with it's precipitate.
D. No, because more CaF_2 can be dissolved in solution.

545. A saturated solution of CaF_2 is created by adding excess CaF_2 to water. The excess CaF_2 precipitate is filtered out of solution. 0.1 *M* HF is added to the solution. Does a precipitate form?

A. Yes, because fluoride ions from the dissociation of HF act as a common ion causing precipitation.
B. Yes, because HF is a strong acid.
C. No, because acid increases the solubility of CaF_2.
D. No, because protons push the equilibrium of the overall reaction to the right balancing out the leftward shift caused by HF.

546. A saturated solution of CaF_2 is created by adding excess CaF_2 to water. The excess CaF_2 precipitate is filtered out of solution. 0.1 *M* HCl is added to the solution. Does a precipitate form?

A. Yes, because chloride ions from the dissociation of HCl act as a common ion causing precipitation.
B. Yes, because HCl is a strong acid.
C. No, because acid increases the solubility of CaF_2.
D. No, because protons push the equilibrium of the overall reaction to the right balancing out the leftward shift caused by HCl.

547. A saturated solution of CaF_2 is created by adding excess CaF_2 to water. The excess CaF_2 precipitate is filtered out of solution. HCl is added to the solution. Next, a small amount of a soluble calcium salt is added to solution. Is a precipitate likely to form?

A. Yes, because the solution was saturated before the addition of the calcium salt.
B. Yes, due to the common ion effect.
C. No, because the solution was not saturated when the calcium salt was added.
D. No, because the solution will become supersaturated.

---

548. What is the equilibrium expression for the dissolution of sucrose (a nonelectrolyte solid) in water?

A. $K = [\text{sucrose}(aq)]$
B. $K = [\text{sucrose}(aq)]/[\text{sucrose}(s)]$
C. $K = [\text{sucrose}(s)][\text{sucrose}(aq)]$
D. $K = [\text{sucrose}(aq)][\text{water}]/[\text{sucrose}(s)]$

549. What is the equilibrium expression for dissolving oxygen in water?

A. $K = [O_2\ (aq)]$
B. $K = [O_2\ (aq)]/[O_2\ (g)]$
C. $K = [O_2\ (g)][O_2\ (aq)]$
D. $K = [O_2\ (aq)][H_2O\ (l)]/[O_2\ (g)]$

550. Which of the following statements is true of an unsaturated solution?

A. If left for long enough, the solution will reach equilibrium.
B. If a catalyst is added, a precipitate will form.
C. If a very small amount of solid is added to the solution, the rate at which that solid dissolves will equal the rate at which new solid is formed.
D. The solution may become saturated if the temperature is changed.

551. Lead (II) chloride has a K_{sp} of 1.6 x 10^{-5}. A 0.001 *M* solution of lead (II) chloride is:

- **A.** Unsaturated
- **B.** Saturated
- **C.** Hypersaturated
- **D.** Supersaturated

552. 100 mL of 0.1 *M* NaCl is mixed with 900 mL of 0.01 *M* $Pb(NO_3)_2$ solution. If the K_{sp} of $PbCl_2$ is 1.6 x 10^{-5}, what is the most likely result?

- **A.** An unsaturated solution
- **B.** A sodium nitrate precipitate
- **C.** A lead (II) chloride precipitate
- **D.** Both sodium nitrate and lead (II) chloride precipitates

553. In which solution will sodium chloride be least soluble?

- **A.** 3.0 *M* hydrochloric acid
- **B.** 5.0 *M* sulfuric acid
- **C.** 3.0 *M* lithium hydroxide
- **D.** Sodium chloride will be equally soluble in all of these solutions

554. How does adding sodium chloride to pure water alter the solubility and K_{sp} of silver chloride in that solution?

- **A.** Both the solubility and K_{sp} are lowered
- **B.** The solubility is lowered, but the K_{sp} remains the same
- **C.** The solubility remains the same, but the K_{sp} is lower
- **D.** Neither is affected

555. How does raising the temperature affect the solubility and K_{sp} of an unknown salt?

- **A.** Both the solubility and the K_{sp} may be affected
- **B.** The solubility may be affected, but the K_{sp} remains the same.
- **C.** The solubility remains the same, but the K_{sp} may be affected.
- **D.** Neither is affected.

556. The K_{sp} of $BaSO_4$, $CaSO_4$, and $PbSO_4$ are $8x10^{-7}$, $9.1x10^{-6}$, and 1.6 x 10^{-8}, respectively. If lithium sulfate were gradually added to a solution containing equal concentrations of barium nitrate, calcium nitrate, and lead (II) nitrate, which precipitate would form first?

- **A.** $BaSO_4$
- **B.** $CaSO_4$
- **C.** $PbSO_4$
- **D.** $LiSO_4$

557. Which expression could be used to find the number of grams of solid NaCl that would have to be added to 20.0 mL of 6.0 *M* $AgNO_3$ to form a saturated solution of silver chloride? Silver chloride has a K_{sp} of 1.8 x 10^{-10}.

- **A.** $\dfrac{(1.8\text{x}10^{-10})(6.0)(20.0)(58.4)}{1000}$
- **B.** $\dfrac{(1.8\text{x}10^{-10})(6.0)(58.4)}{(20)(1000)}$
- **C.** $\dfrac{(1.8\text{x}10^{-10})(6.0)(20.0)(58.4)}{(20)(6.0)(1000)}$
- **D.** $\dfrac{(1.8\text{x}10^{-10})(20.0)(58.4)}{(6.0)(1000)}$

558. Iodine, which is only slightly soluble in pure water, dissolves readily in 1 *M* sodium iodide solution. Which of the following is a possible explanation for this observation?

- **A.** The presence of iodide ions in solution enhances the solubility of iodine
- **B.** Iodide is a powerful oxidizing agent: it reduces the iodine, bringing it into solution
- **C.** The iodine reacts with the iodide ions, thus keeping the concentration of iodine in the solution low
- **D.** The sodium ions balance the charge of the newly formed iodide ions

Solubility Factors

559. Which of the following could change an unsaturated solution of silver chloride into a saturated solution?

- **I.** Lowering the temperature
- **II.** Adding sodium chloride
- **III.** Evaporating some of the water

- **A.** I only
- **B.** II only
- **C.** I and II only
- **D.** I, II, and III

560. How does increasing the partial pressure of a gas affect its solubility in water?

- **A.** The solubility of the gas decreases
- **B.** The solubility of the gas increases
- **C.** The solubility of the gas remains the same
- **D.** The effect on the solubility of the gas depends on the gas

561. When cold water is placed in a glass, small bubbles can eventually be seen to form on the inside of the glass. Which of the following is the most likely explanation for this observation?

A. As the water warms up, oxygen from the air is absorbed into the water.
B. As the water warms up, oxygen in the water is released.
C. As the glass cools down, oxygen in the water condenses on to the sides of the glass.
D. As the glass cools down, oxygen in the glass is released into the water.

562. A certain carbonated beverage is stored in pressurized cans. Which of the following might cause gas to escape from the beverage?

I. lowering the pressure above the beverage by opening the can
II. vigorously shaking the can
III. heating the can
IV. putting salt into the beverage after it is opened

A. I and II only
B. II and III only
C. I, II, and III only
D. I, II, III, and IV

563. With few exceptions, oxygen and sunlight are required to support animal life. Based upon this fact and the solubility of gases in water at different temperatures, in what part of the ocean is the greatest biomass found?

A. in the warm waters of the coral reefs
B. at the bottom of the sea
C. in the cold waters of the arctic
D. coastal regions just north and south of the equator

Use the equation below to answer questions 564-568. The equation below gives the relationship between the partial pressure of a gas P_g over a solution and its solubility χ_g (here given as the mole fraction of gas dissolved in solution). k is Henry's law constant and varies with the type of gas and the solvent. The equation is known as Henry's law.

$$\chi_g = kP_g$$

564. According to Henry's law, if the partial of a gas over a solution is doubled, its solubility is:

A. decreased by a factor of 2.
B. remains the same.
C. increased by a factor of 2.
D. increased by a factor of 4.

565. Henry's law holds for a slightly soluble gas as long as the gas:

A. has a high molecular weight.
B. has a low molecular weight.
C. is diatomic.
D. does not react chemically with the solvent.

566. The Henry's law constants for N_2 and O_2 in water are $8.6\text{x}10^{-4}$ /kPa and $4.4\text{x}10^{-4}$ /kPa respectively. There are approximately 10^5 Pa in 1 atm. Air is approximately 79% nitrogen and 21% oxygen. What is the approximate ratio of nitrogen to oxygen molecules dissolved in a glass of water that has been left exposed to 1 atm of air pressure for a long time?

A. 1:2
B. 2:1
C. 4:1
D. 8:1

567. $4.4\text{x}10^{-4}$ /kPa is the Henry's law constant for O_2 in water. There are approximately 10^5 Pa in 1 atm. If the air is approximately 20% oxygen. What is the mole fraction of oxygen in a glass of water that has reached equilibrium with air at 1 atm?

A. $8.8\text{x}10^{-3}$
B. $8.8\text{x}10^{-2}$
C. $8.8\text{x}10^{-1}$
D. 8.8

568. The graph below shows the mole fraction of hydrogen and nitrogen gas in water at 50 °C and various pressures. The slope of the dotted line is equal to Henry's law constant for the respective gases.

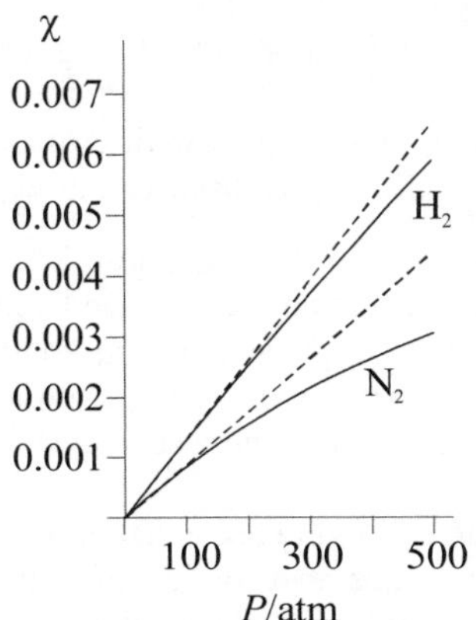

Deviations in Henry's law probably occur:

A. at low pressures.
B. at high pressures.
C. with light gasses.
D. with heavy gasses.

Phases

569. Which of the following most accurately describes the movement of molecules in a solid?

A. held perfectly motionless by strong bonds with other molecules or atoms
B. vibrating in place close to other molecules or atoms
C. moving closely past each other in random directions vibrating, rotating, and forming and breaking bonds with other molecules or atoms
D. moving past each other at a distance in random directions vibrating and rotating, without bonding

570. Which of the following most accurately describes the movement of molecules in a liquid?

A. held perfectly motionless by strong bonds with other molecules or atoms
B. vibrating in place close to other molecules or atoms
C. moving closely past each other in random directions vibrating, rotating, and forming and breaking bonds with other molecules or atoms
D. moving past each other at a distance in random directions vibrating and rotating, without bonding

571. Which of the following most accurately describes the movement of molecules in a gas?

A. held perfectly motionless by strong bonds with other molecules or atoms
B. vibrating in place close to other molecules or atoms
C. moving closely past each other in random directions vibrating, rotating, and forming and breaking bonds with other molecules or atoms
D. moving past each other at a distance in random directions vibrating and rotating, without bonding

572. Which of the following best describes the process by which carbon dioxide is dissolved in water?

A. Carbon dioxide changes phase from gas to liquid
B. Carbon dioxide changes phase from gas to aqueous
C. Water changes phase from liquid to aqueous
D. Carbon dioxide undergoes a chemical reaction with the water to form a new substance

573. Which of the following are constant throughout a single phase of any substance or mixture at equilibrium not separated by rigid or adiabatic walls?

I. Temperature
II. Pressure
III. Mole Fraction

A. I only
B. II only
C. III only
D. I, II, and III

574. Pentanol is not completely soluble in water. When pentanol and water are mixed, two layers form; one has a small amount of water dissolved in pentanol, and the other a small amount of pentanol dissolved in water. How many phases are present?

A. None; there are no pure substances in this system
B. One; the entire system is liquid
C. Two; each layer is a distinct phase
D. Four: the pentanol dissolved in water, the water as solvent, the water dissolved in pentanol, and the pentanol as solvent

575. When hard chocolate is heated, it gradually becomes softer and softer, eventually flowing like a viscous liquid. Which of the following is NOT a true statement?

A. Melting chocolate is an endothermic process
B. The temperature of chocolate remains constant as it is transformed from solid to liquid
C. When chocolate is heated, the additional kinetic energy allows the constituent molecules to move past each other more freely
D. For chocolate, solid and liquid cannot be considered distinct phases

576. A glass of ice water sits in a room. The room, the water in the glass, and the ice are at 0 °C. How many phases of water are there in the room?

A. one
B. two
C. three
D. The situation as described is physically impossible.

577. Many long-chain unsaturated hydrocarbons are liquids at room temperature, but when the molecules are hydrogenated, they become solids. Which of the following is the most likely explanation for this?

A. The hydrogenation enhances the amount of hydrogen bonding present
B. The sigma-bonds in the hydrogenated compound rotate more easily than the pi-bonds in the unsaturated compound, allowing the molecules to pack together more easily
C. The hydrogens form dimers, doubling the size of the molecules
D. The heat of hydrogenation provides the energy for the phase change

578. A container is partially filled with a mixture of ether and water. The space above the mixture is vacuumed and the container is sealed. The container is shaken and then allowed to sit. Which of the following is true after several hours have passed?

A. Both the liquid portion and the gas portion of the container contain two phases.
B. Both the liquid portion and the gas portion of the container contain one phase.
C. The liquid portion has two phases and the gas portion has one phase.
D. The liquid portion has two phases and there is no gas portion.

Heat Capacity

579. Equal amounts of heat are added to 10-gram blocks of metal X and metal Y. The temperature of metal X is found to increase by twice as much as that of metal Y. Which of the following is the most likely explanation for this observation?

A. The specific heat of metal X is twice as great as that of metal Y.
B. The specific heat of metal X is half as great as that of metal Y.
C. The density of metal X is twice as great as that of metal Y.
D. The density of metal X is half as great as that of metal Y.

580. Which of the following is the amount of energy that 1 g of a substance must absorb in order to change its temperature by 1 K?

A. heat capacity
B. specific heat
C. molar heat capacity
D. volume heat capacity

581. Water has a higher heat capacity than many other molecules because:

A. water molecules are light and thus able to move faster to absorb more heat energy.
B. the atoms of water molecules are able to absorb a greater proportion of heat energy via vibration.
C. the hydrogen bonds of water break to absorb heat energy.
D. the hydrogen bonds of water form to absorb heat energy.

582. Copper has a specific heat of 0.24 cal/g K. How much heat is required to raise the temperature of 10 grams of copper from 20°C to 25°C?

A. 1.2 cal
B. 12 cal
C. 60 cal
D. 667 cal

583. Benzene has a specific heat of 0.25 cal/g°C. If 20 calories of heat is added to 10 grams of benzene at 12°C, what is the new temperature of the benzene?

A. 12.5°C
B. 17°C
C. 20°C
D. 62°C

584. What is the heat capacity of a 15-gram sample of water?

A. 0.015 cal/K
B. 0.07 cal/K
C. 4.18 cal/K
D. 15 cal/K

585. What is the molar heat capacity of water?

A. 2.0 cal/mol K
B. 18.0 cal/mol K
C. 55.0 cal/mol K
D. 100 cal/mol K

586. The human body is 60% water. The specific heat of water allows the body to:

I. stay warm longer in cold environments.
II. cool off more quickly in hot environments.
III. warm slowly in hot environments.

A. I only
B. I and II only
C. I and III only
D. I, II, and III

587. Ethanol (molecular weight 46) has a molar heat capacity of 111 J/mol K. How much heat is required to raise the temperature of 3 moles of ethanol by 5°C?

A. 6 J
B. 36 J
C. 1.7 kJ
D. 76.6 kJ

588. Which of the following is true when calculating the change in energy of a substance using the equation: $Q = mc\Delta T$?

A. Work done on the substance does not need to be considered when using this equation.
B. *c* is the specific heat of the substance.
C. *m* is the molality of the substance.
D. ΔT is the change in temperature of the substance in kelvins.

589. What is the specific heat of water at 100 °C and 1 atm?

A. Zero, because there is no change in temperature as heat is added.
B. 1 cal/g °C
C. Infinite, because there is no change in temperature as heat is added.
D. It depends upon the amount of water.

Use the diagram below to answer questions 590-593. The diagram below shows a cross-section of an insulated cylinder-piston apparatus filled with gas and connected to a heat reservoir. The volume of the cylinder can be expanded via the piston at constant pressure or constant temperature.

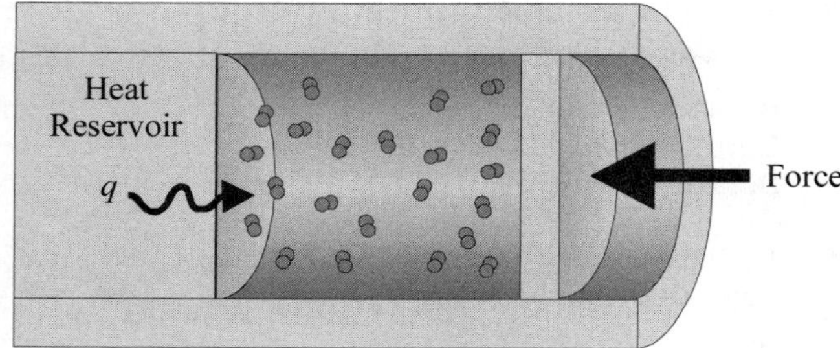

590. Which of the following is true concerning the gas molecules in the cylinder?

I. If no heat is added to the gas, it will cool as it expands.
II. The average kinetic energy of the molecules is proportional to the temperature of the gas.
III. The gas does work on the surroundings as it expands.

A. I only
B. I and II only
C. II and III only
D. I, II, and III

591. If the cylinder is held at constant volume and heat is added, what will become of the added energy?

A. The energy does work on the surroundings.
B. If heat is added, the gas must expand.
C. All the energy becomes kinetic energy of the gas molecules raising the temperature and pressure of the gas.
D. All the energy becomes kinetic energy of the gas molecules raising the temperature but not the pressure of the gas.

592. If the piston were held still and the gas in the cylinder were replace with a gas having a greater heat capacity, which of the following would be true as heat was added to the gas?

A. The pressure increase would be greater for the same amount of heat added.
B. The pressure increase would be less for the same amount of heat added.
C. The temperature increase would be greater for the same amount of heat added.
D. The temperature increase would be the same for the same amount of heat added.

593. If the cylinder is allowed to expand while heat is added so that the pressure stays constant, what will become of the added heat energy?

A. All the energy does work on the surroundings.
B. All the energy becomes kinetic energy of the gas molecules raising the temperature of the gas.
C. Some of the energy does work on the surroundings and some of the energy becomes the kinetic energy of the gas molecules raising the temperature of the gas.
D. The cylinder cannot be expanded at constant pressure because the added heat energy becomes molecular kinetic energy which must raise the temperature and the pressure.

---

594. Does the same gas have a greater heat capacity at constant pressure or constant volume?

A. Constant pressure because energy is required both to do work and raise the temperature.
B. Constant pressure because work is done on the gas as the piston moves.
C. Constant volume because all the energy goes into increasing the kinetic energy of the molecules.
D. Constant volume because the increase in pressure uses some of the heat energy.

595. Liquids and solids don't expand as much as gases so the work that they do on the surroundings when expanding is not usually significant in terms of their heat capacity. Yet, the constant pressure heat capacity for solids and liquids is greater than the constant volume heat capacity for solids and liquids. Why?

A. Since solids and liquids are less compressible, they are able to exert a greater force over the small distance doing more work on the surroundings.
B. Since solids and liquids are less compressible, they are able to exert a greater force over the small distance doing more work on the solid or liquid.
C. Molecular attractions between the molecules of solids and liquids are significant, and a small increase in intermolecular distance represents a large decrease in intermolecular potential energy.
D. Molecular attractions between the molecules of solids and liquids are significant, and a small increase in intermolecular distance represents a large increase in intermolecular potential energy.

596. Larger objects made from the same substance have:

A. greater specific heats.
B. smaller specific heats.
C. the same specific heats.
D. Whether or not the specific heat increases or decreases for different amounts of the same substance depends upon the substance.

597. Larger objects made from the same substance have:

A. greater heat capacities.
B. smaller heat capacities.
C. the same heat capacities.
D. Whether or not the heat capacity increases or decreases for different amounts of the same substance depends upon the substance.

598. Is it possible for a substance to have a negative heat capacity?

A. No, because a negative heat capacity would indicate that the temperature of the substance would decrease when energy is added.
B. No, because a negative heat capacity would indicate that an endothermic process cools the surroundings.
C. Yes, some solids have negative heat capacities because they cool other objects when they make contact.
D. Yes some gases have negative heat capacities because they can expand and cool while being heated.

Use the table below to answer questions 599-601. The table below gives the specific heats of several elements and compounds at 298 K.

Substance	Specific Heat (J/g·K)		Substance	Specific Heat (J/g·K)
$N_2(g)$	1.04		$H_2O(l)$	4.18
$Al(s)$	0.90		$H_2O(s)$	2.03
$Cu(s)$	0.38		$H_2O(g)$	1.84
$Au(s)$	0.13		$CH_4(g)$	2.20
$Ar(g)$	0.52		$CO_2(g)$	0.84

599. Three cold metal bars of equal weight, surface area, and temperature are placed in the warm sunlight. The objects are made of aluminum, copper, and gold, respectively. Which of the following is true? (Assume that the difference in emissivity for each surface is not a factor.)

A. The temperature of the gold bar will increase the fastest.
B. The temperature of the aluminum bar will increase the fastest.
C. The gold bar will reach the highest temperature.
D. The aluminum bar will reach the highest temperature.

600. Which phase of water can absorb the most energy while changing its temperature the least?

A. ice
B. liquid water
C. steam
D. All are equal.

601. Which of the following might explain why methane has a higher specific heat than Argon?

A. The structure of argon allows it more ways to absorb energy.
B. The structure of methane allows it more ways to absorb energy.
C. Methane molecules move faster than argon molecules at a given temperature.
D. Argon is heavier.

---

602. Which of the following compounds is likely to have the greatest molar heat at 298 K?

A. C
B. CH_4
C. C_2H_6
D. C_3H_8

603. 20.0 grams of Liquid A at 5°C are added to a fully insulated beaker containing 65.0 grams of Liquid B at 25°C. No chemical reaction occurs. If the temperature in the beaker is 15°C when the liquids reach thermal equilibrium, what conclusion can reasonably be drawn?

A. Liquid A has a higher specific heat than liquid B
B. Liquid B has a higher specific heat than liquid A.
C. Liquid A has a negative specific heat.
D. Liquid B has a negative specific heat.

Calorimeters

604. The heat transfer in a coffee cup calorimeter corresponds to:

A. the enthalpy change.
B. the entropy change.
C. the Gibbs free energy change.
D. the energy change.

605. Reactions within a coffee cup calorimeter are considered to take place at constant pressure for calculation purposes. Why is constant pressure a good approximation in this case?

A. The vapor pressure of solution is approximately equal to atmospheric pressure.
B. There is very little fluid pressure because the solution is shallow. This makes the pressure very close to 1 atm throughout the solution.
C. The reaction usually occurs too fast for a pressure change to be significant.
D. The volume of the solution increases or decreases with temperature, holding the pressure constant.

606. A dilute aqueous solution of HCl is added to a dilute aqueous solution of NaOH in a coffee cup calorimeter. In order to approximate the enthalpy change for this reaction using $mc\Delta T$, what value should the student give to c?

A. The student may use the specific heat of pure water.
B. The student must look up the specific heat of NaCl solution at a specific concentration.
C. The student must use the heat of formation of water.
D. The student must use the heat of formation of NaCl.

607. A stirring mechanism, such as a straw, typically accompanies a coffee cup calorimeter. The function of the stirring mechanism is to ensure:

A. a heterogeneous solution.
B. a single phase.
C. constant pressure.
D. solution formation by adding energy.

608. A student mixed 50 ml of 1 *M* HCl and 50 ml of 1 *M* NaOH in a coffee cup calorimeter and calculated the change in enthalpy to be −2.7 kJ. He next tried the same experiment with 100 ml of 1 *M* HCl and 100 ml of 1 *M* NaOH. The calculated enthalpy change was:

A. −1.4 kJ
B. −2.7 kJ
C. −5.2 kJ
D. −10.4 kJ

609. A student mixed 50 ml of 1 *M* HCl and 50 ml of 1 *M* NaOH in a coffee cup calorimeter and calculated the molar heat of solution to be −54 kJ/mol. He next tried the same experiment with 100 ml of 1 *M* HCl and 100 ml of 1 *M* NaOH. The calculated the molar heat of solution for his second trial was:

A. −27 kJ/mol
B. −54 kJ/mol
C. −108 kJ/mol
D. −216 kJ/mol

610. 50.0 grams of water at 80°C are added to 150.0 grams of water at 25°C. If the temperatures are allowed to equilibrate with no heat loss to the environment, what is the final temperature of the water?

A. 1.25 °C
B. 39 °C
C. 66 °C
D. 100 °C

611. 30.0 grams of liquid A at 5°C are added to 55.0 grams of liquid B at 30°C. If the result is a single compound at a temperature of 35°C, what conclusion can reasonably be drawn?

A. Liquid A is more stable than liquid B
B. Liquid B is more stable than liquid A
C. Heat was lost to the environment
D. Liquid A and liquid B underwent an exothermic reaction

612. Water and sodium chloride are mixed within a bomb calorimeter. The experimenter then calculates $mc\Delta T$ for the resulting solution. If the calorimeter is well-insulated, the result should reflect:

A. the energy of solution
B. the free energy of solution
C. the heat capacity of solution
D. the free entropy of solution

613. An ideal coffee cup calorimeter is an example of a(n):

A. isolated system.
B. closed system.
C. open system.
D. free system.

614. An ideal bomb calorimeter is an example of a(n):

A. Isolated system
B. Closed system
C. Open system
D. Free system

615. The heat transfer in a bomb calorimeter corresponds to:

A. the enthalpy change.
B. the entropy change.
C. the Gibbs free energy change.
D. the energy change.

616. Two solutions at room temperature are mixed in a coffee cup calorimeter and an exothermic reaction occurs. As the temperature of the calorimeter rises, heat transfers:

A. from the surroundings to the solution.
B. from the solution to the surroundings.
C. No heat is transferred between the surroundings and the solution.
D. The direction of heat transfer depends upon room temperature.

617. If C is the heat capacity of a bomb calorimeter and ΔT is the temperature increase during a reaction, which of the following expression gives the heat transfer?

A. $C/\Delta T$
B. $C\Delta T$
C. $mC\Delta T$
D. m $C/\Delta T$

618. Suppose the combustion of 1 g of benzoic acid, $C_7H_6O_2$, produces 26 kJ of energy. In a certain bomb calorimeter, a 10 °C temperature increase is observed when 2 g of benzoic acid are combusted. What is the heat capacity of the calorimeter?

A. 1.3 kJ/°C
B. 2.6 kJ/°C
C. 5.2 kJ/°C
D. 10.4 kJ/°C

619. The combustion of hydrazine, N_2H_4, produces nitrogen gas and water vapor. The heat of combustion for this reaction is −618 kJ/mol. If 1.6 g or hydrazine are combusted in a bomb calorimeter at 298 K and with a heat capacity of 6.2 kJ/°C, what will be the temperature of the bomb calorimeter after the reaction?

A. 288 K
B. 293 K
C. 303 K
D. 308 K

Phase Changes

620. When water melts, what kind(s) of bonds are breaking?

I. Covalent
II. Ionic
III. Hydrogen

A. I only
B. III only
C. I and II only
D. I and III only

621. When sodium chloride melts, what kind(s) of bonds are breaking?

I. Covalent
II. Ionic
III. Dipole-dipole

A. I only
B. II only
C. III only
D. II and III

622. Put the following compounds in order of boiling point from highest to lowest?

I. H_2O
II. NH_3
III. CH_4

A. I, II, III
B. I, III, II
C. II, III, I
D. III, II, I

623. When dry ice (CO_2) melts, what kind of bonds are broken?

A. Ionic
B. Covalent
C. Intermolecular
D. Metallic

624. When sodium hydrogen phosphate (Na_2HPO_4) melts, what kind of bonds are broken?

A. Ionic
B. Covalent
C. Intermolecular
D. Metallic

625. When diamond melts, what kind of bonds are broken?

A. Ionic
B. Covalent
C. Intermolecular
D. Metallic

626. When copper melts, what kind of bonds are broken?

A. Ionic
B. Covalent
C. Intermolecular
D. Metallic

627. Which of the following compounds should have the highest boiling point?

A. BrCl
B. Br_2
C. ICl
D. Xe

628. When mass spectrometry is performed on a sample of hydrogen fluoride gas, a large peak is seen at 20 amu, with successively smaller peaks at 40 amu, 60 amu, 80 amu, and 100 amu. When hydrogen chloride is used instead, only one peak at 36.5 amu is seen. What is an explanation for this?

A. The hydrogen fluoride dissociates at high energies
B. Hydrogen bonding allows the HF molecules to form small associations in the vapor phase
C. Hydrogen chloride is ionic while hydrogen fluoride is covalent
D. Hydrogen chloride is a liquid under standard conditions, while hydrogen fluoride is a gas

629. If a solid at a temperature below its melting point is heated, the added energy:

I. breaks bonds.
II. increases the translational kinetic energy of the molecules.
III. increases the vibrational kinetic energy of the molecules.

A. I only
B. II only
C. III only
D. I, II, and III

630. If a solid at its melting point is heated, the added energy:

I. breaks bonds.
II. increases the translational kinetic energy of the molecules.
III. increases the vibrational kinetic energy of the molecules.

A. I only
B. II only
C. III only
D. I and II only

631. If a liquid at a temperature below its boiling point is heated, the added energy:

I. breaks bonds.
II. increases the translational kinetic energy of the molecules.
III. increases the vibrational kinetic energy of the molecules.

A. I only
B. III only
C. I and II only
D. II and III only

632. If a liquid at its boiling point is heated, the added energy:

I. breaks bonds.
II. increases the translational kinetic energy of the molecules.
III. increases the vibrational kinetic energy of the molecules.

A. I only
B. II only
C. III only
D. I and II only

633. Which of the following gives an approximate value for the heat of sublimation of a substance?

A. Add the heat of vaporization and the heat of fusion
B. Subtract the heat of fusion from the heat of vaporization
C. Multiply the heat of formation by the heat of vaporization
D. Divide the heat of reaction by the heat of formation

634. Which of the following physical processes when occurring in a pure substance is accompanied by an increase in temperature?

I. melting
II. vaporization
III. sublimation

A. I only
B. I and II only
C. I, II, and III
D. None are accompanied by a temperature increase.

635. 3 grams of solid H_2O are heated to 0°C. If, just after this temperature is reached, 6 more calories are added, the temperature of the H_2O will be:

A. 0 °C
B. 2 °C
C. 3 °C
D. 6 °C

636. Water has a heat of fusion of 80 cal/g and a heat of vaporization of 540 cal/g. How much heat is required to convert 100 grams of ice at 0°C to steam at 100°C?

A. (80 + 540) cal
B. 100(80 + 540) cal
C. (100)(1)(1000) cal
D. 100(80 + 100 + 540) cal

637. Water has a heat of fusion of 80 cal/g and a heat of vaporization of 540 cal/g. If it takes 10 minutes to heat a pot of water from 20°C to 100°C, and the heat is delivered at a constant rate, how much longer does it take for all the water to be converted to steam?

A. Less than 1 minute
B. Between 1 minute and ten minutes
C. Between ten minutes and one hour
D. More than one hour

638. When a sample of water is evaporating at a temperature below its boiling point, what happens to the temperature of the liquid phase?

A. It tends to cool, since the most rapidly moving molecules are leaving the liquid phase
B. It tends to cool, since the transition from liquid to vapor is exothermic
C. It tends to heat, since the transition from liquid to vapor is exothermic
D. It tends to heat, since the gas phase has a higher entropy than the liquid phase

639. A beaker containing an unknown liquid is placed in a liquid nitrogen bath. The temperature drops rapidly to −35°C, then rises to −32°C. It remains at −32°C for some time, and then drops again, rapidly at first, and then more slowly. Finally, the temperature stabilizes at −196°C. Which of the following can be concluded from this data?

A. The freezing point of the liquid is −32°C
B. The freezing point of the liquid is −35°C
C. The freezing point of the liquid is −196°C
D. The freezing point of the nitrogen is −32°C

640. Food can be preserved by removing water via sublimation in a process called freeze-drying. Freeze drying probably occurs at:

A. low temperatures and pressures.
B. low temperatures and high pressures.
C. high temperatures and low pressures.
D. high temperatures and pressures.

Questions 641-647 utilize this heating curve for 10 grams of an unknown substance:

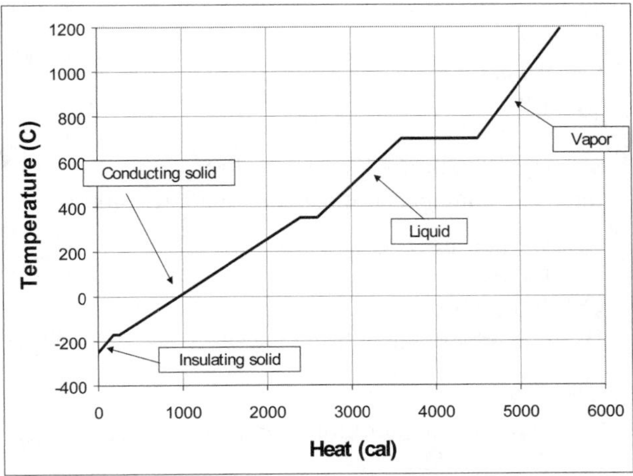

641. Which phase of this substance has the greatest specific heat?

A. Insulating solid
B. Conducting solid
C. Liquid
D. Vapor

642. What is the melting point of this substance?

A. −170°C
B. 350°C
C. 2400°C
D. Cannot be determined from this graph

643. If the pressure exerted on the liquid is increased, the melting point:

A. will increase.
B. will decrease.
C. will remain the same.
D. cannot be determined from the information given.

644. If the pressure of the gas above the fluid is increased, the boiling point of the substance will:

A. Increase
B. Decrease
C. Remain the same
D. Cannot be determined from the information given

645. What is the latent heat of fusion of this substance?

A. 20 cal/g
B. 240 cal/g
C. 260 cal/g
D. Cannot be determined from this graph

646. What is the specific heat of the vapor phase of this substance?

A. 0.08 cal/g°C
B. 0.2 cal/g°C
C. 0.5 cal/g°C
D. 1.0 cal/g°C

647. Starting at −200°C, how much heat would have to be removed from the substance to bring it to −400°C?

A. 50 calories
B. 100 calories
C. 200 calories
D. It is impossible to bring the substance to −400°C

---

648. Ice is just beginning to form on a winter lake. Where is the lake the coldest?

A. At the top of the lake
B. At the bottom of the lake
C. At the middle depths of the lake
D. The lake is the same temperature everywhere.

Use the diagram and formula below to answer questions 649-651. The diagram below shows the $\ln(P_{vapor})$ vs. $1/T$ for four substances. The Clausius-Clapeyron equation describes this relationship.

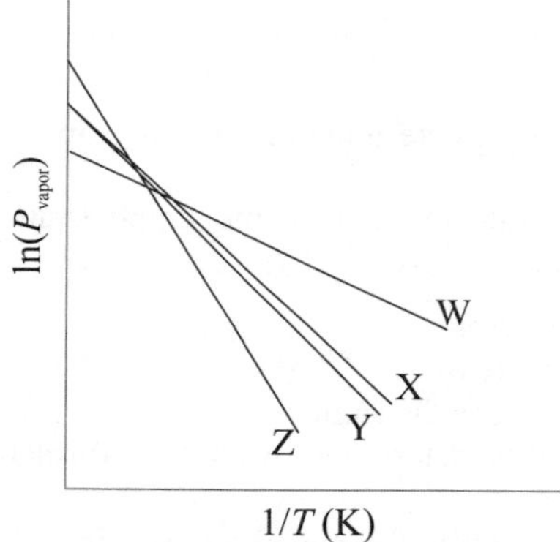

$$\ln(P_{vap}) = -\frac{\Delta H_{Vap}}{R}\left(\frac{1}{T}\right) + C$$

The Clausius-Clapeyron Equation

649. According to the graph, at high temperatures, which substance has the greatest vapor pressure?

A. W
B. X
C. Y
D. Z

650. Which substance has the greatest heat of vaporization?

A. W
B. X
C. Y
D. Z

651. If X is water, which line could represent diethyl ether, $(C_2H_5)_2O$?

A. W
B. X
C. Y
D. Z

Phase Diagrams

Use the phase diagram below to answer questions 652-662.

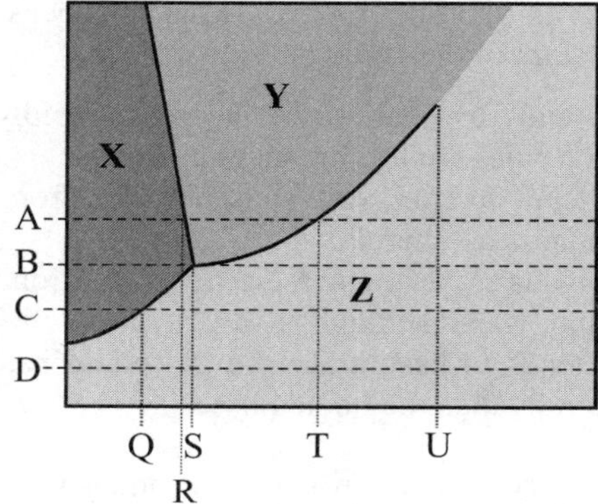

652. From the diagram, which of the following represents solid, liquid, and gas respectively?

A. X, Y, Z
B. Y, X, Z
C. Z, X, Y
D. X, Z, Y

653. If the phase diagram is of water, which line might represent one atmosphere of pressure?

A. A
B. B
C. C
D. D

654. If the phase diagram is of water, which line could represent 100 °C?

A. R
B. S
C. T
D. U

655. If the phase diagram is of water, and line A is one atm, which line could represent 0 °C?

A. Q
B. R
C. S
D. T

656. If the phase diagram is of water, and line A is one atm, can ice, liquid water, and water vapor exist in equilibrium at 0 °C and 1 atm?

A. No, because the triple point is below 1 atm.
B. No, because the triple point is below 0 °C.
C. Yes, because ice and water can have an equilibrium with the vapor phase regardless of the pressure or temperature.
D. Yes, because 1 atm and 0 °C is above the triple point.

657. Why might this phase diagram resemble water?

A. There are three phases represented.
B. The equilibrium line between solid and liquid slopes up and to the left.
C. The equilibrium line between liquid and gas slopes up and to the right.
D. The equilibrium line between solid and gas slopes down and to the left.

658. Which line goes through the triple point?

A. Q
B. R
C. S
D. U

659. At constant pressure C, what does the substance do as temperature increases at Q?

A. condense
B. sublime
C. deposit
D. melt

660. If the substance is held at a constant temperature S, and pressure is increased from very low to very high, the substance:

A. condenses
B. deposits
C. evaporates
D. freezes

661. Which of the following represents the critical temperature?

A. Q
B. R
C. S
D. U

662. If a small amount of salt were added to the substance forming a solution, which of the following could represent the phase diagram of the solution? (The dark lines represent the equilibrium lines of the pure substance and the dashed lines represent the equilibrium lines of the solution.)

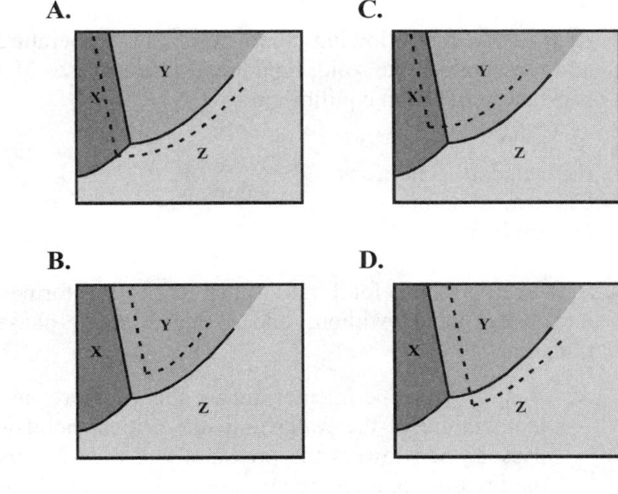

---

663. A substance at a temperature and pressure above its critical temperature and pressure is called:

A. plasma
B. gas
C. supercritical fluid
D. hyperphased

664. Which of the following is NOT true for supercritical fluid?

A. It will not become a liquid no matter how great the pressure.
B. It will not become a liquid no matter how low the temperature.
C. The liquid and vapor phases are indistinguishable.
D. It is found at temperatures and pressures above the critical temperature and critical pressure.

665. A substance is dissolved in supercritical CO_2. If the solution is allowed to equilibrate at room temperature and pressure, the CO_2 will:

A. boil out of solution.
B. precipitate out of solution.
C. condense and remain in solution.
D. condense and separate out of solution.

666. On a diagram of pressure versus temperature, how does the transition between liquid and gas phases appear?

A. As a point
B. As a straight line segment
C. As a quadrilateral region
D. As a curved line segment

667. At which of the following combinations of temperature and pressure can the solid, liquid, and vapor phases of a substance coexist in equilibrium?

A. STP
B. standard conditions
C. critical point
D. triple point

668. Is it ever possible for liquid water to be transformed into water vapor without undergoing a sharp phase transition?

A. Yes, if an experimenter raises the pressure and temperature of the water until the critical point is passed, then lowers the pressure, and then lowers the pressure and temperature until the vapor phase is reached.
B. Yes, if an experimenter lowers the temperature of the water until it freezes, and then allows it to undergo sublimation to the vapor phase.
C. No, it is never possible to change from liquid to vapor without a sharp phase transition.
D. No, it is possible for amorphous substances to change from liquid to vapor without a sharp phase transition, but water is not an amorphous substance.

Use the following information and the diagram below to answer questions 669-676. Water and butanol are *partially miscible.* At room temperature, when equal amounts of water and butanol are poured into the same container, two phases are formed, one containing mostly water with a small amount of butanol, and one containing mostly butanol with a small amount of water. At some critical solution temperature water and butanol are *completely miscible* and only one phase exists for any mixture. The miscibility graph below shows temperature vs. mole fraction for a hypothetical Fluid X with water at a constant pressure of 1 atm. The shaded region represents the two phase system and the light region represents the one phase system.

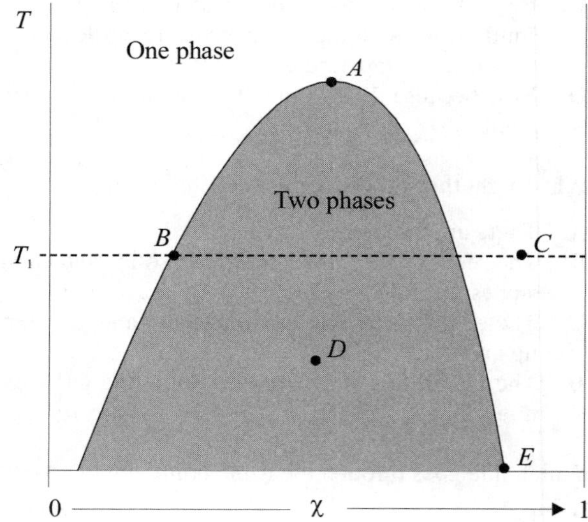

669. Which point on the diagram is at the critical solution temperature?

A. *A*
B. *B*
C. *D*
D. *E*

670. What does the system look like at point *C*?

A. A small amount of Fluid X dissolved in water
B. A small amount of water dissolved in Fluid X
C. A homogeneous mixture of equal amounts of Fluid X and water.
D. A two phase system of water and Fluid X with more Fluid X than water.

671. What does the system look like at point D?

A. A heterogeneous mixture of a small amount of Fluid X dissolved in water in one phase and a small amount of water dissolved in Fluid X in the other phase.
B. A two phase system with both phases containing an equal amount of water and Fluid X.
C. A homogeneous mixture of equal amounts of Fluid X and water.
D. A two phase system of water and Fluid X with more Fluid X than water.

672. A student pours 100 mL of pure water into a 500 mL beaker. The student adds 10 mL of Fluid X stirs and allows the mixture to come to equilibrium. The student repeats this process until the beaker is full. If the entire experiment is performed at temperature T_1, which of the following describes the system during this experiment?

A. Initially Fluid X dissolves completely in water to form a single phase. When enough Fluid X is added, a second phase of water dissolved in Fluid X begins to form. As the amount of Fluid X increases, the water from the first phase is dissolved into the Fluid X to form a single phase.
B. A two phase system with Fluid X in one phase and water in the other phase remain separate throughout the experiment.
C. Initially Fluid X dissolves completely in water to form a single phase. As more Fluid X is added, the Fluid X begins to predominate so that water is dissolved in Fluid X. One phase is maintained throughout.
D. Initially Fluid X dissolves completely in water to form a single phase. As the solution nears the critical solution temperature, equal amounts of Fluid X and water form two separate phases. As the amount of Fluid X increases still further, the water from the first phase is dissolved into the Fluid X to form a single phase.

673. Fluid X is probably:

A. nonpolar.
B. polar, but less polar than water.
C. exactly as polar as water.
D. more polar than water.

674. Water and triethylamine are only *partially miscible* at high temperatures, but become *completely miscible* at low temperatures. Which of the following might represent the miscibility graph for water and triethylamine?

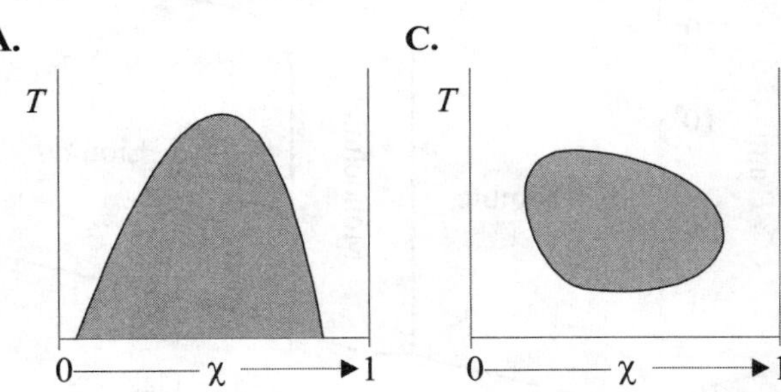

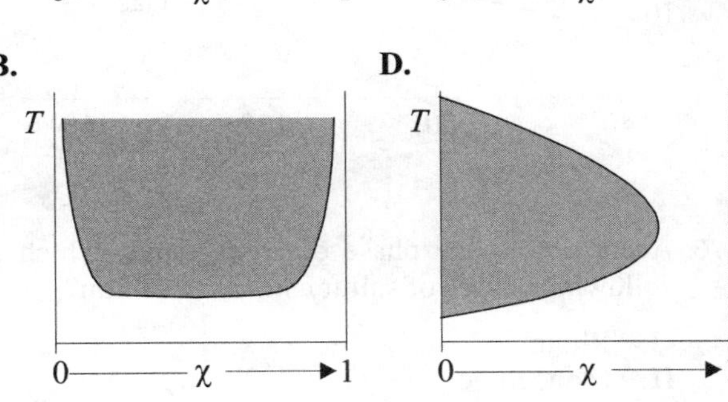

675. Below is the miscibility graph for water and butanol. What is the critical solution temperature for water and butanol?

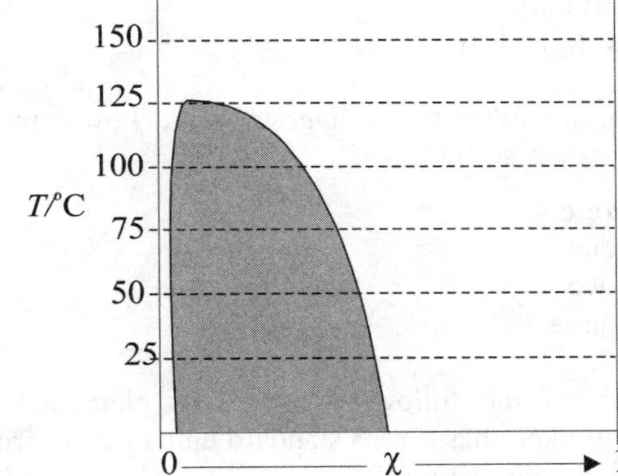

A. 25 °C
B. 50 °C
C. 125 °C
D. 150 °C

Use the diagram below to answer questions 676-681. Below is a phase diagram for elemental sulfur.

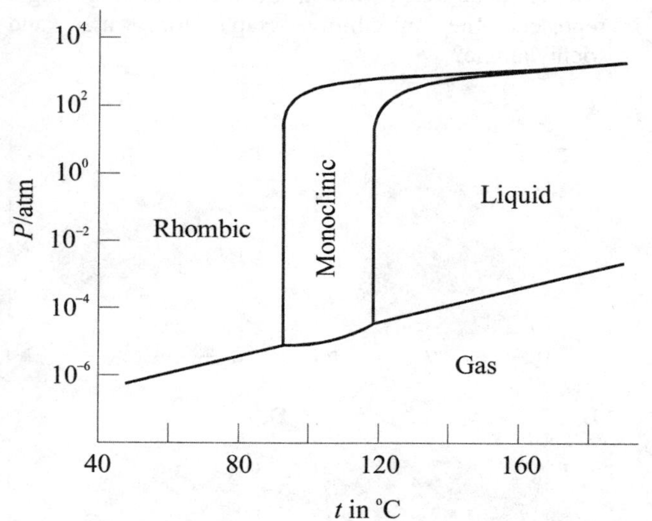

676. According to the phase diagram above, which of the following phases of sulfur can exist at 1 atm?

I. Rhombic
II. Monoclinic
III. Liquid

A. I only
B. II only
C. III only
D. I, II, and III

677. According to the phase diagram above, how many triple points does sulfur have?

A. zero
B. one
C. two
D. three

678. Which of the following phases of elemental sulfur mostly likely has a zero standard enthalpy of formation at room temperature?

A. Rhombic
B. Monoclinic
C. Liquid
D. Gas

679. A sample of sulfur is initially at 1 atm and 60 °C. The pressure on the sample is decreased to 10^{-6} atm. Next the temperature is increased slowly to 90 °C. According to the phase diagram, which of the following are the phase changes experienced by the sample?

A. sublimation
B. vaporization
C. transition and vaporization
D. transition and condensation

680. A sample of sulfur is initially at 10^{-6} atm and 70 °C. If the pressure on the sample is increased, what phase change will the sample experience?

A. sublimation
B. deposition
C. vaporization
D. fusion

681. The Frosch process is used to mine sulfur. In the Frosch process water is pumped into a sulfur deposit at 1 atm to melt the sulfur. Compressed air then forces the sulfur to the surface where it is cooled and collected. Which of the following must true in order for this process to occur?

A. The sulfur deposit must exist as monoclinic sulfur.
B. The water used to melt the sulfur must be superheated.
C. The compressed air must be at a pressure less than one atmosphere.
D. The sulfur is ultimately collected as a gas.

Use the diagram below to answer questions 682-689. The isotherms of H_2O are shown as solid lines on pressure vs. molar volume phase diagram below.

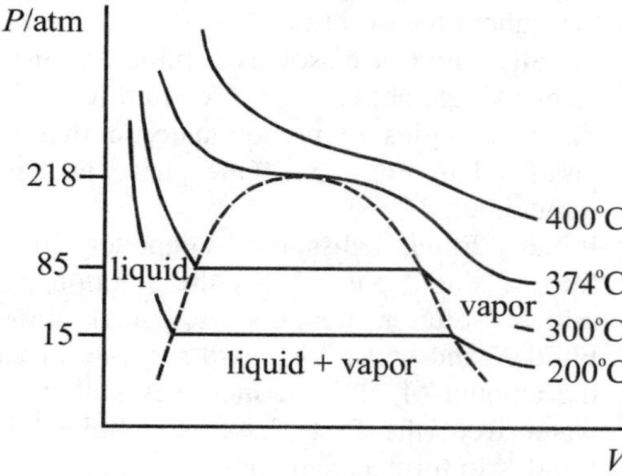

682. What are the critical pressure and temperature of water?

A. 15 atm and 200 °C
B. 85 atm and 300 °C
C. 218 atm and 374 °C
D. above 218 atm and 400 °C

683. Which of the following describes the volume change that occurs along the horizontal section of the 200 °C isotherm moving from left to right?

A. The volume decreases as the pressure goes up.
B. The volume is decreases as water condenses.
C. The volume increases as the pressure goes down.
D. The volume increases as water is vaporized.

684. What is the boiling point of water at 85 atm?

- **A.** 200 °C
- **B.** 300 °C
- **C.** 374 °C
- **D.** 400 °C

685. What is the minimum pressure required to convert H_2O to liquid water at 400 °C?

- **A.** less than 218 atm
- **B.** 218 atm
- **C.** Greater than 218 atm
- **D.** H_2O will not turn to liquid at 400 °C no matter how great the pressure.

686. At which of the following temperatures is liquid H_2O impossible?

- **A.** 200 °C
- **B.** 374 °C
- **C.** 400 °C
- **D.** H_2O liquid can exist at any temperature if the pressure is high enough.

687. What is the name for H_2O at 219 atm and 375 °C?

- **A.** liquid
- **B.** vapor
- **C.** supercritical fluid
- **D.** ice IX

688. As we follow the 200 °C isotherm from left to right, which of the following describes the energy flow? (Assume ideal behavior for water vapor.)

- **A.** The net energy flow is zero along the entire path.
- **B.** The net energy flow is into the sample along the entire path.
- **C.** The net energy flow is out of the sample along the entire path.
- **D.** The net energy flow is into the sample along the horizontal portion of the path, and zero at all other points.

689. As we follow the 200 °C isotherm from left to right before it enters the liquid and vapor area, which of the following describes the energy flow?

- **A.** Work is done *on* the sample and heat is *released* from the sample.
- **B.** Work is done *on* the sample and heat is *absorbed* by the sample.
- **C.** Work is done *by* the sample and heat is *released* from the sample.
- **D.** Work is done *by* the sample and heat is *absorbed* by the sample.

Colligative Properties

690. Colligative properties depend upon:

- **A.** quality not quantity.
- **B.** size not charge.
- **C.** number not kind.
- **D.** mass not size.

691. Which of the following is NOT a colligative property?

- **A.** Boiling point elevation
- **B.** Osmotic pressure
- **C.** Vapor pressure lowering
- **D.** Heat of solution

692. Which of the following is true when a nonvolatile solute is added to a pure liquid?

- **A.** The melting and boiling points decrease.
- **B.** The melting and boiling points increase.
- **C.** The melting point increases and the boiling point decreases.
- **D.** The melting point decreases and the boiling point increases.

693. Why does the boiling point rise when a nonvolatile solute is added to a pure liquid?

- **A.** Because the density of the solution becomes greater than the density of the pure liquid.
- **B.** Because the kinetic energy of the molecules of the liquid in the solution is less than the kinetic energy of the molecules in the pure liquid.
- **C.** Because adding a solute to a liquid cools the liquid.
- **D.** Because solute molecules occupy surface area of the liquid and lower the vapor pressure.

694. Why does the freezing point drop when a nonvolatile solute is added to a pure liquid?

- **A.** Because the solute disrupts the formation of a symmetrical crystalline structure.
- **B.** Because the kinetic energy of the molecules of the liquid in the solution is less than the kinetic energy of the molecules in the pure liquid.
- **C.** Because adding a solute to a liquid warms the liquid.
- **D.** Because solute molecules occupy surface area of the liquid and lower the vapor pressure.

695. Farmers who have to leave fruit in unheated storage rooms sometimes put the fruit next to large containers of water to keep the fruit from freezing. Why would this method work?

A. The water moistens the fruit, and the higher water content of the fruit decreases its freezing temperature, protecting the fruit.
B. The water absorbs humidity from the environment, drying out the fruit, and the fruit can then become cold without danger of ice crystals forming.
C. Because the water in fruit contains dissolved sugar, the fruit freezes at a higher temperature than the water, so although the fruit does freeze, it is less seriously damaged.
D. Because the water in fruit contains dissolved sugar, the container of water freezes at a higher temperature than the fruit. The freezing water releases heat, which prevents the temperature in the room from dropping until all the water is frozen.

696. Which of the following aqueous solutions would have the lowest freezing point?

A. 0.10 M $CaCl_2$
B. 0.45 M $NaNO_3$
C. 0.65 M HCl
D. 0.90 M NH_3

697. The freezing-point depression constant of water is 1.86 K/m. What concentration of sodium chloride is necessary to cause a container of water to freeze at −5°C?

A. 1.3 m
B. 1.86 m
C. 3.0 m
D. 5.0 m

698. Cyclohexane has a freezing point of 6.5°C and a freezing-point depression constant of 20.1 K/m. If a 0.01 m solution of acetic acid is dissolved in the cyclohexane, what is the freezing point of the mixture? (Acetic acid does not dissociate in cyclohexane.)

A. 0°C
B. 0.2°C
C. 4.8°C
D. 6.3°C

699. Acetone has a boiling point of 56.2°C and a boiling-point elevation constant of 1.71 K/m. If 2.0 grams of dimethyl ether (C_2H_6O) are dissolved in 100 grams of acetone, what is the boiling point of the resulting solution? (Assume that dimethyl ether does not dissociate in acetone.)

A. 56.2°C
B. 56.9°C
C. 59.6°C
D. 90.2°C

700. 20 grams of an unknown non-volatile substance is dissolved in 1.0 kg of a solvent, and the resulting freezing point of the solvent is measured and compared to the freezing point of the pure solvent. In order to calculate the molecular weight of the unknown solvent, what else must be known?

I. The freezing-point depression constant of the solvent
II. The degree to which the solute ionizes in the solvent
III. The density of the resulting solution

A. I only
B. II only
C. I and II only
D. I, II, and III

701. When 10 grams of a certain alkane is dissolved in 1 kg of benzene, the freezing point is found to be 4.9°C. Which of the following is most likely to be the alkane? (Benzene has a freezing point of 5.5°C and a freezing-point depression constant of 5.1 K/m.)

A. Ethane (C_2H_6)
B. Butane (C_4H_{10})
C. Hexane (C_6H_{14})
D. Octane (C_8H_{18})

702. When 10 grams of a certain material are dissolved in 1 kg of water, the freezing point of the solution is −0.46°C. Which of the following could be the substance? (Water has a freezing-point depression constant of 1.86 K/m.)

A. $HC_2H_3O_2$
B. HNO_3
C. NaCl
D. $CaCl_2$

703. According to the equation $\Delta T = K_f mi$, the freezing point of 0.1 M NaCl solution should be −0.372 °C. However, the experiment gives a value of −0.348 °C. Which of the following might explain this discrepancy?

A. Precipitation results in fewer particles in solution.
B. Ion pairing between Na^+ and Cl^- ions results in fewer particles in solution.
C. Under real conditions impurities in the water create more particles.
D. Water auto ionizes creating more particles in solution.

704. The van't Hoff factor found by experiment is often less than the theoretical van't Hoff factor. If ΔT_c is the change in temperature calculated assuming no dissociation and ΔT_o is the change in temperature found by experiment, which of the following might give the van't Hoff factor found by experiment?

A. $\Delta T_c \Delta T_o$
B. $1/(\Delta T_c \Delta T_o)$
C. $\Delta T_c/\Delta T_o$
D. $\Delta T_o/\Delta T_c$

705. The observed van't Hoff factor is often found to be less than the theoretical van't Hoff factor. Is this discrepancy increased or decreased when the solution is diluted?

A. Decreased, because there is more ion pairing
B. Decreased, because there is less ion pairing.
C. Increased, because there is more ion pairing.
D. Increased, because there is less ion pairing.

706. A container filled with water is divided in half by a semipermeable membrane. When salt is placed in one half of the container, the water level on one side of the container rises above the level on the other side. Which of the following is an accurate description of this process?

A. Since evaporation occurs more readily in the side with the salt, that side will have the lower water level.
B. Since the concentration of water is higher on the pure water side, more water will flow to that side, and the water level of the salt side will be lower.
C. Since the concentration of water is higher on the pure water side, more water will flow from that side to the other side, and the water level of the salt side will be higher.
D. It is impossible for an unequal distribution of water to occur; if water can flow from one side to another, the levels will always equalize.

707. If a solution with high osmotic pressure is separated from a solution with low osmotic pressure by a membrane permeable only to water:

A. water will tend to move into the solution with high osmotic pressure.
B. water will tend to move into the solution with low osmotic pressure.
C. solute will tend to move into the solution with high osmotic pressure.
D. solute will tend to move into the solution with low osmotic pressure.

708. A swimming pool is filled with pure water and the hydrostatic pressure at a certain depth is measured at 2.5×10^4 Pa. Salt is added to the water in the pool. After the salt dissolves, the pressure will be equal to:

A. the hydrostatic pressure plus the osmotic pressure.
B. the hydrostatic pressure plus the osmotic pressure.
C. the hydrostatic pressure only.
D. the osmotic pressure only.

709. 0.01 mole of glucose is added to one liter of water at 25 °C and 1 atm. The osmotic pressure is 0.24 atm. What is the total pressure in the container?

A. 0.24 atm
B. 0.76 atm
C. 1 atm
D. 1.24 atm

710. When placed in a hypertonic solution, a red blood cell undergoes crenation (it shrivels). The cell undergoes crenation because:

A. the osmotic pressure inside the cell is *less* than the osmotic pressure outside the cell.
B. the osmotic pressure inside the cell is *greater* than the osmotic pressure outside the cell.
C. the hydrostatic pressure inside the cell is *less* than the hydrostatic pressure outside the cell.
D. the hydrostatic pressure inside the cell is *greater* than the hydrostatic pressure outside the cell.

711. A U-shaped tube with a semi-permeable membrane at its lowest point is filled with water, as shown below. (The water heights as indicated in the diagram represent initial conditions.) If the left half of the tube contains glucose at a concentration of 3 molar, which of the following expressions could be used to estimate the distance, in meters, that the water on the left side rises? (The ideal gas law constant is 8.314 J/mol K. Assume that the temperature is 27°C.)

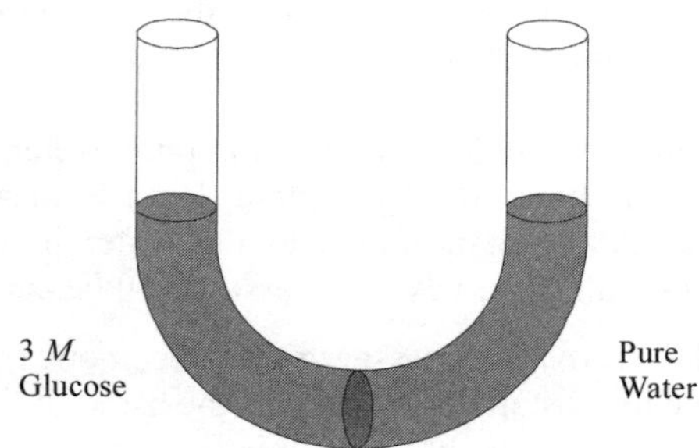

A. $\frac{(8.3)(27)}{(3)(1000)(9.8)}$

B. $\frac{(3)(8.3)(27)}{(1000)(9.8)}$

C. $\frac{(3)(8.3)(300)}{(1000)(9.8)}$

D. $\frac{(8.3)(300)}{(3)(1000)(9.8)}$

712. Suppose a membrane were found which was permeable to ions but not to isobutanol. If this membrane were used to separate a solution of NaCl in isobutanol from pure isobutanol, which side would reach the higher level?

A. The salt side, due to osmotic pressure
B. The pure isobutanol side, since the salt will flow out but not in
C. Neither side, since the solvent cannot pass through the membrane.
D. It would depend on whether the volume of the pure solvent increased or decreased when salt was added to it.

Use the following information to answer questions 713-715. Orderly systems possess more free energy than disorderly systems. In solutions solute molecules disrupt the natural order of a pure liquid. Osmotic potential is the free energy of a solution under constant temperature and pressure. Pure water is arbitrarily assigned an osmotic potential value of zero. Water spontaneously flows from high to low osmotic potential.

713. Based on the information above, the osmotic potential of an aqueous solution of NaCl:

A. is negative
B. is positive
C. is zero
D. may be either negative or positive.

714. The decrease in the osmotic potential of water molecules is directly proportional to the osmotic concentration. Which of the following solutions has the lowest osmotic potential?

A. 0.9 *M* NaCl
B. 0.9 *M* glucose
C. 1 *M* NaCl
D. 1 *M* glucose

715. Unlike osmotic potential, water potential is a function of temperature and pressure as well as solute concentration. When two solutions are separated by a membrane permeable to water but not solute, water will move from higher water potential to lower water potential. As the hydrostatic pressure of a solution increases, the water potential:

A. decreases because the water molecules have less space and so become more ordered.
B. remains constant because water has no place to move.
C. remains constant, because temperature increases.
D. increases because the water molecules have more free energy.

Definitions

716. The definition of pH is:

A. $\log[H^+]$
B. $-\log[H^+]$
C. 7
D. the number of hydrogen ions in solution.

717. Which of the following defines a neutral solution?

A. pH = 7
B. $[H^+] = [OH^-]$
C. $[H^+] = 0$
D. A solution where acids don't donate protons and bases don't accept protons.

718. The autoionization of water (shown below) is an endothermic reaction.

$$2H_2O \rightarrow H_3O^+ + OH^-$$

as the temperature of pure water increases the pH:

A. decreases because $[H^+]$ increases.
B. increases because $[OH^-]$ increases.
C. increases because $[H^+]$ increases.
D. remains at 7 because $[H^+]$ equals $[OH^-]$ even after an equilibrium shift.

719. Which of the following is true of the equilibrium for the autoionization of water (shown below)?

$$2H_2O \rightarrow H_3O^+ + OH^-$$

A. The equilibrium lies a little to the left.
B. The equilibrium lies far to the left.
C. The equilibrium lies far to the right.
D. The equilibrium is balanced evenly between the left and the right.

720. All of the following are true concerning acids and bases EXCEPT:

A. acids taste sour.
B. bases taste bitter.
C. bases are slippery.
D. acids are sticky.

721. Which of the following is NOT a way of defining an Arrhenius acid?

A. An Arrhenius acid is a substance that increases the H^+ concentration in aqueous solution
B. An Arrhenius acid is a substance that increases the pH of an aqueous solution
C. An Arrhenius acid is a substance that increases the pOH of an aqueous solution
D. An Arrhenius acid is a substance that decreases the OH^- concentration in an aqueous solution

722. Which of the following is/are Arrhenius acids?

I. HCl
II. CH_3CH_2OH
III. BF_3

A. I only
B. III only
C. I and II only
D. I, II, and III

723. Which of the following can act as Bronsted-Lowry acids in some circumstances?

I. HCl
II. CH_3CH_2OH
III. BF_3

A. I only
B. III only
C. I and II only
D. I, II, and III

724. Which of the following can act as Lewis acids in some circumstances?

I. HCl
II. CH_3CH_2OH
III. BF_3

A. I only
B. III only
C. I and II only
D. I, II, and III

725. In the following reaction, which substance is acting as a Brönsted-Lowry acid?

$$NH_3 + H^- \rightarrow NH^{2-} + H_2$$

A. NH_3
B. H^-
C. NH_4^+
D. NH_2^-

726. Which of the following is NOT possible?

A. A Lewis acid that is not a Brönsted-Lowry acid
B. A Brönsted-Lowry acid that is not a Lewis acid
C. A substance that is sometimes a Lewis acid and sometimes a Lewis base
D. A substance that is sometimes a Brönsted-Lowry acid and sometimes a Brönsted-Lowry base

727. A warning sign in an industrial laboratory says "Do not store acids with bases! Acids go in cabinet A, bases in cabinet B." The definition of acid and base that this sign is using is most likely:

A. the Arrhenius definition, since unlike the Brönsted-Lowry and Lewis definitions, it is not possible for a substance to be sometimes an acid and sometimes a base.
B. the Brönsted-Lowry definition, since it is the definition most commonly used.
C. the Lewis definition, since it involves reactions between substances.
D. the Lewis definition, since it can be applied to almost all substances.

728. BF_3 is a very effective Lewis acid. Which of the following is a reasonable explanation for this observation?

A. BF_3 contains fluorine
B. BF_3 is negatively charged
C. BF_3 has an empty orbital in its valence shell
D. BF_3 holds its protons very loosely

729. What is the pH of a solution with a hydrogen ion concentration of 1.0×10^{-4} *M*?

A. –4
B. –0.0001
C. 0.0001
D. 4

730. What is the pH of a solution with a hydrogen ion concentration of 3.0×10^{-4} *M*?

A. 3.0
B. 3.5
C. 4.0
D. 4.5

731. What is the hydrogen ion concentration in a solution with pH 11.26?

A. 5.5×10^{-12}
B. 5.5×10^{-11}
C. 5.5×10^{-10}
D. 1.12×10^{-6}

732. 1 mL of a strong acid solution has a pH of 2.3. It is diluted with water to make a 100-mL solution. What is the new pH?

A. 0.023
B. 2.3
C. 4.3
D. 230

733. What is the conjugate base of ammonia?

A. NH_2^-
B. NH_3
C. NH_4^+
D. OH^-

734. What is the conjugate base of HCO_3^-?

A. H_2CO_3
B. CO_3^{2-}
C. H_2O
D. OH^-

735. What is the conjugate base of H_3O^+?

A. H_2O
B. H_3O
C. OH^-
D. H_2O_2

736. What is the conjugate acid of phosphoric acid?

A. $H_2PO_4^-$
B. H_3PO_4
C. $H_4PO_4^+$
D. H_3O^+

737. Which of the following conducts electricity the most poorly?

A. NaCl(l)
B. NaCl(aq)
C. HCl(aq)
D. HClO(aq)

738. Which of the following is the weakest acid?

A. Hydrochloric acid
B. Sulfuric acid
C. Nitric acid
D. Formic acid

739. Which of the following is amphiprotic in aqueous solution?

I. HCl
II. HPO_4^{2-}
III. H_2O

A. I only
B. II only
C. I and II only
D. II and III only

740. Which of the following is NOT true concerning conjugate bases?

A. Every acid has a conjugate base.
B. The conjugate base of a Brönsted-Lowry acid is that acid minus its acidic proton.
C. Some conjugate bases are acidic.
D. OH^- is the conjugate base of H_3O^+.

741. Which of the following is most useful for calculating the pH of a solution?

A. The strength of the conjugate base
B. The K_a
C. The concentration of hydrogen ions
D. The concentration of the acid

742. Which of the following is NOT true?

A. Strong acids have weak conjugate bases.
B. Weak acids have strong conjugate bases.
C. The weaker the acid, the stronger its conjugate base.
D. The stronger the acid, the weaker its conjugate base.

743. If HA is a stronger acid than H_3O^+:

A. HA will transfer its proton to H_2O more effectively than H_3O^+ will transfer its proton to A^-.
B. H_3O^+ will transfer its proton to H_2O more effectively than HA will transfer its proton to A^-.
C. HA will transfer its proton to A^- more effectively than H_3O^+ will transfer its proton to H_2O.
D. HA will transfer its proton to H_3O^+ more effectively than H_2O will transfer its proton to A^-.

744. Which of the following is the weakest acid?

A. $HClO_4$
B. H_3O^+
C. HBr
D. HNO_3

745. Which of the following is the strongest acid?

A. HF
B. HCl
C. HBr
D. HI

746. Which of the following is the weakest base?

A. H^-
B. Na_2O
C. N^{3-}
D. OH^-

747. Which of the following is the weakest base in aqueous solution?

A. KOH
B. $Ca(OH)_2$
C. $Mg(OH)_2$
D. OH^-

748. Place the following in order from strongest to weakest base?

I. ClO^-
II. H_2O
III. Cl^-

A. I, II, III
B. I, III, II
C. II, I, III
D. III, I, II

749. Which of the following factors determine the percent ionization of an acid?

I. temperature of solution
II. identity of acid
III. concentration of acid

A. II only
B. I and II only
C. II and III only
D. I, II, and III

750. Which of the following is true concerning percent dissociation of an acid?

A. Percent dissociation increases with concentration.
B. Percent dissociation is 100% for strong acids regardless of concentration.
C. Percent dissociation typically decreases as temperature increases.
D. Percent dissociation is greater for stronger acids.

751. In biological systems, many reactions involve proton transfers. What is one ramification of this observation?

A. Most biological reactions occur under low pH conditions.
B. Most biological reactions occur under high pH conditions.
C. In living organisms, small fluctuations in pH can create large rate changes.
D. In living organisms, large fluctuations in pH are required to create small rate changes.

752. Which of the following expressions gives the pH of a buffered solution?

A. $\log[H^+]$
B. $-\log[H^+]$
C. $pK_a - \log([A^-]/[HA])$
D. $pK_b + \log([A^-]/[HA])$

753. If the pH is increased by one pH point, the proton concentration is:

A. decreased by a factor of two.
B. decreased by a factor of ten.
C. increased by a factor of two.
D. increased by a factor of ten.

754. Compared to a solution with a pH of 5, how many H^+ ions are in a solution with a pH of 3?

A. 4 times as many
B. 10 times as many
C. 20 times as many
D. 100 times as many

755. Alcohols are weaker acids than water. Pure water has a pH of 7. A pH greater than 7 indicates a base. How can alcohols be weaker acids than water without being bases?

A. Alcohols are slightly basic.
B. A pH lower than 7 doesn't always indicate a base.
C. Alcohols are less willing to give up their acidic hydrogen to a base than is water.
D. Water is slightly acidic.

Factors Determining Acid Strength

756. In a series of oxyacids, the acidity increases:

A. as the number of hydrogens attached to the central atom decreases.
B. as the number of hydrogens attached to the central atom increases.
C. as the oxidation number of the central atom decreases.
D. as the oxidation number of the central atom increases.

757. Which of the following oxides forms a base when dissolved in water?

A. K_2O
B. CO_2
C. SO_2
D. NO_2

758. Which of the following oxyacids is the strongest acid in aqueous solution?

A. $HClO$
B. $HClO_2$
C. $HClO_3$
D. $HClO_4$

759. Rank the following oxyacids from strongest acid to weakest acid in aqueous solution?

I. $HClO$
II. $HBrO$
III. HIO

A. I, II, III
B. II, III, I
C. III, II, I
D. III, I, II

760. Why are oxyacids with more oxygens around the central atom stronger acids?

A. Because each oxygen can take on a proton
B. Because oxygens are electron withdrawing and can neutralize hydroxide ions.
C. Because oxygens around the central atom strengthen the bond between the oxygen and the acidic hydrogen.
D. Because oxygens around the central atom withdraw electrons increasing the polarity of the bond between the oxygen and the acidic hydrogen.

761. H represents the acidic proton on conjugate base X^-. Which of the following tends to decrease the acidity of HX:

A. a polar H–X bond.
B. a strong H–X bond.
C. a stable conjugate base X^-.
D. a high temperature.

Hydrides

762. What is a hydride?

A. A metal atom without a hydration shell
B. Compounds that have been formed by the removal of water.
C. Compounds containing only two elements where one of the elements is hydrogen
D. Compounds containing hydrogen.

763. Which of the following molecular hydrides has the highest boiling point?

A. SiH_4
B. H_2Te
C. H_2S
D. SnH_4

764. On the periodic table, the acidity of hydrides increases moving from:

I. left to right.
II. right to left.
III. top to bottom.

A. I only
B. II only
C. I and III only
D. II and III only

765. Which of the following is generally true concerning hydrides (other than NH_3)?

A. All hydrides are *acidic* or neutral.
B. All hydrides are *basic* or neutral.
C. Metal hydrides are *acidic* or neutral, while nonmetal hydrides are *basic* or neutral.
D. Metal hydrides are *basic* or neutral, while nonmetal hydrides are *acidic* or neutral.

766. Which of the following is the strongest acid?

A. H_2O
B. H_2S
C. H_2Se
D. H_2Te

767. Which of the following is NOT a strong base in aqueous solution?

A. LiH
B. CaH_2
C. H_2
D. KH

768. Which of the following statements is true?

I. NaH is a stronger base than NaOH
II. CaH_2 is a stronger base than $Ca(OH)_2$
III. HCl is a stonger acid than HClO

A. I only
B. I and II only
C. I and III only
D. I, II, and III

769. Which of the following is a true statement concerning the reaction:

$$H^-(aq) + H_2O(l) \rightleftharpoons H_2(g) + OH^-(aq)$$

A. The equilibrium lies to the *left* because OH^- is a stronger base than H^-.
B. The equilibrium lies to the *left* because H^- is a stronger base than OH^-.
C. The equilibrium lies to the *right* because OH^- is a stronger base than H^-.
D. The equilibrium lies to the *right* because H^- is a stronger base than OH^-.

Equilibrium Constants for Acid-Base Reactions

770. Hydrofluoric acid is a weak acid ($pK_a = 3.1$). In terms of its ionization in water, fluoride is:

A. a weak acid
B. a weak base
C. a strong acid
D. a strong base

771. What is the acid dissociation constant K_a of water?

A. 1.0×10^{-14}
B. 1.8×10^{-16}
C. 1
D. 10

772. A glass of pure water sits at room temperature. Which of the following is true?

A. There is about one hydrated H^+ for every one billion water molecules in the glass.
B. There are roughly equal numbers of hydrated H^+ ions and water molecules in the glass.
C. There are roughly equal numbers of hydrated OH^- ions and water molecules in the glass.
D. The total number of hydrated H+ ions and OH^- ions exceeds the total number of water molecules.

773. Which of the following explains why the measured pH of a solution represents an equilibrium condition?

A. The ΔG of an acid-base reaction is negative.
B. The ΔH of an acid-base reaction is negative.
C. The ΔS of an acid-base reaction is positive.
D. Proton transfer reactions take place very fast.

774. Given the pK_a table below, which of the following choices is the strongest base?

Acid	pK_a
HSCN	−1.8
HBF_4	0.5
HIO	10.5

A. SCN^-
B. BF_4^-
C. IO^-
D. Cl^-

775. Which of the following pK_a's would indicate that the associated acid is a weak one?

I. –3.0
II. 4.5
III. 11.2

A. I only
B. II only
C. I and II only
D. II and III only

776. Given that the molarity of pure water is 55.6 *M*, what is the pK_a of water?

A. 7.0
B. 14.0
C. 15.7
D. 55.6

777. Acetic acid has a pK_a of 4.74. What is the pK_b of the acetate ion?

A. 2.26
B. 4.74
C. 9.26
D. Cannot be determined from the information given

778. Hydrocyanic acid has a pK_a of 9.31. What is the pK_b of the cyanide ion?

A. 2.31
B. 4.69
C. 9.31
D. Cannot be determined from the information given

779. The pK_a of the amphoteric hydrogen carbonate ion is 10.25. What is the pK_b of this ion?

A. 3.25
B. 3.75
C. 10.25
D. Cannot be determined from the information given

780. Benzoic acid is a weak acid with a pK_a of 4.19. Its conjugate, the benzoate ion, is:

A. A strong acid
B. A strong base
C. A weak acid
D. A weak base

781. Which of the following techniques could be used to find the pK_a of a weak acid?

I. Measure the amount of base needed to neutralize the acid
II. Measure the pH of a given concentration of the acid
III. Measure the pH halfway to the equivalence point of a titration
IV. Find the pK_b of its conjugate base

A. I only
B. II and III only
C. II, III, and IV only
D. I, II, III, and IV

782. Consider the reaction:

$$HCN + C_2H_3O_2^- \rightleftharpoons CN^- + HC_2H_3O_2$$

The pK_a of HCN is 9.31 and the pK_a of $HC_2H_3O_2$ is 4.74. If the reaction is initiated under standard conditions, what is true about the relative concentrations of HCN and $HC_2H_3O_2$ once the reaction reaches equilibrium?

A. There is less HCN than $HC_2H_3O_2$
B. There are equal amounts of HCN and $HC_2H_3O_2$
C. There is more HCN than $HC_2H_3O_2$
D. More information is needed in order to know which acid is dominant

Finding pH

783. What is the pH of a 0.01 *M* solution of HBr?

A. 0.01
B. 1.0
C. 2.0
D. 12.0

784. What is the pH of a 0.01 *M* solution of KOH?

A. 0.01
B. 1.0
C. 2.0
D. 12.0

785. What is the pOH of a 0.01 *M* solution of KOH?

A. 0.01
B. 1.0
C. 2.0
D. 12.0

786. What is the pH of a 0.1 *M* solution of acetic acid? (pK_a = 4.74)

A. 1.0
B. 2.9
C. 11.1
D. 13.0

787. What is the pH of a 0.1 *M* solution of hydrocyanic acid? (pK_a= 9.3)

A. 5.1
B. 8.3
C. 9.3
D. 10.3

788. What is the pH of a 7 *M* HCl solution?

A. –0.85
B. 0.85
C. 1.15
D. 1.85

789. An acid is added to water to make 30 mL of a solution with a pH of 6.5. If the solution is then diluted to a total volume of 300 mL, what will be the new value of the pH?

A. 6.5
B. Between 6.5 and 7
C. 7.5 because the hydrogen ion concentration is reduced by a factor of 10.
D. Either A or B depending upon what acid was used, but C is impossible.

790. What is the pH of a 10^{-8} *M* HCl solution?

A. 6
B. 6.96
C. 7
D. 8

791. Solution A is composed of aqueous hydrofluoric acid (pK_a=3.74). Solution B is composed of aqueous acetic acid (pK_a=4.74). Which solution has the higher pH?

A. Solution A
B. Solution B
C. The pH of the two solutions is the same
D. Cannot be determined without more information

792. Which of the following influence the pH of an aqueous solution of a weak base?

I. Concentration of the base
II. pK_b of the base
III. Volume of the solution

A. I only
B. II only
C. III only
D. I and II only

793. In a 1 *M* solution of acetic acid (pK_a=4.7), the vast majority of the acetic acid is undissociated. Aspartic acid residues in proteins, which have an extremely similar molecular structure to acetic acid, are generally almost completely dissociated. Why?

A. In aspartic acid, nearby hydrophobic residues activate the acid.
B. Proteins are generally found in extremely acidic environments, increasing the dissociation of residues such as aspartic acid.
C. Proteins contain basic residues stronger than water which deprotonate the aspartic acid.
D. The many acidic residues found in a typical protein increase the effective concentration to above 1 *M*.

794. Substance X is amphiprotic with pK_a = 5.7. Which of the following is true about the pH of a 3.0 *M* solution of substance X?

A. The pH is less than 7
B. The pH is 7
C. The pH is more than 7
D. It is impossible to determine anything about the pH without more information

795. A 50 mL solution of nitric acid has a pH of 1.3. If the solution is diluted to a total volume of 5.0 L, what is the new pH?

A. 0.3
B. 1.3
C. 2.3
D. 3.3

796. Hydroiodic acid has a much higher K_a than hydrochloric acid. In aqueous solution, however, equal concentrations of each produce essentially the same pH. Which of the following is the most reasonable explanation for this observation?

A. Hydroiodic acid is less soluble in water than hydrochloric acid.
B. Both acids are much weaker than water, so essentially no dissociation takes place.
C. Both acids are much stronger than water, so dissociation is essentially 100% in both cases.
D. Both acids are much stronger than hydronium, so dissociation is essentially 100% in both cases.

797. Hydroiodic acid has a much higher K_a than hydrochloric acid. In aqueous solution equal concentrations of each produce essentially the same pH. However, in acetic acid solution, Hydroiodic acid produces a lower pH. Which of the following is the most reasonable explanation for this observation?

A. Hydroiodic acid is a stronger acid than acetic acid.
B. The conjugate base of hydrochloric acid is a stronger base than water.
C. Acetic acid is a weaker base than water.
D. Acetic acid is a stronger base than water.

798. Propanol has a much higher K_a than ethane. In aqueous solution, however, equal concentrations of each produce essentially the same pH. Which of the following is a reasonable explanation for this observation?

A. Propanol is less soluble in water than ethane
B. Both acids are much weaker than water, so essentially no dissociation takes place
C. Both acids are much stronger than water, so dissociation is essentially 100% in both cases
D. Both acids are much stronger than hydronium, so dissociation is essentially 100% in both cases

Salts

799. Hydrochloric acid is a strong acid. A solution of sodium chloride should have a pH of:

A. less than seven
B. seven
C. more than seven but less than fourteen
D. more than fourteen

800. A strong acid and a weak base are mixed to produce salt and water. The salt is dried and added to pure water. The pH of the salt solution is:

A. less than seven.
B. seven.
C. greater than seven.
D. Whether the pH is above or below seven depends upon the concentration of the salt.

801. Which of the following cations is a weak acid in aqueous solution?

A. Sr^{2+}
B. Ca^{2+}
C. Ba^{2+}
D. Mg^{2+}

802. Which of the following cations is a weak acid in aqueous solution?

A. Na^{+}
B. Al^{3+}
C. Cs^{+}
D. Ca^{2+}

803. Which of the following salts does not act as an acid in aqueous solution?

A. $Zn(NO_3)_2$
B. $Cr(NO_3)_3$
C. $NaNO_3$
D. $Al(NO_3)_3$

804. What is happening in the following reaction?

$$Fe(H_2O)_6^{3+}(aq) \rightleftharpoons Fe(H_2O)_5(OH)^{2+}(aq) + H^+(aq)$$

A. An iron ion is hydrated.
B. A hydrated iron ion acts as an acid causing one of the water molecules in its hydration shell to release a proton.
C. A hydrated iron ion acts as a base creating one hydroxide ion in its hydration shell.
D. Water is hydrolyzed by an iron ion.

805. Which of the following cations is the strongest acid in aqueous solution?

A. Fe^{3+}
B. Ca^{2+}
C. Cu^{2+}
D. Zn^{2+}

806. Sodium acetate is:

A. A weak base and a weak electrolyte
B. A weak base and a strong electrolyte
C. A strong base and a weak electrolyte
D. A strong base and a strong electrolyte

807. Which of the following salts is (are) basic?

I. $NaNO_3$
II. NaCN
III. Na_2CO_3

A. None of the above
B. I only
C. II and III only
D. I, II, and III

808. The K_a of NH_4^+ is 5.6×10^{-10}. The K_b of CN^- is 2×10^{-5}. The pH of an aqueous solution of the salt NH_4CN is:

A. below 7 because NH_4^+ is more acidic than CN^- is basic.
B. below 7 because NH_4^+ is more acidic than CN^-.
C. above 7 because CN^- is more basic than NH_4^+ is acidic.
D. above 7 because CN^- is more basic than NH_4^+.

809. Which of the following salts will be less soluble in acidic solutions than in basic solutions?

A. NaOH
B. NaCl
C. KCl
D. NH_4Cl

810. Which of the following is the strongest acid?

A. $Mg_3(PO_4)_2$
B. $CaHPO_4$
C. K_2HPO_4
D. NaH_2PO_4

811. Consider the polyprotic acid H_2CO_3 (pK_{a1}=6.37, pK_{a2}=10.25). An aqueous solution of $NaHCO_3$ would have a pH of:

A. less than seven.
B. seven.
C. greater than seven.
D. An unknown value; whether the solution is acidic or basic would depend on the concentration of the salt.

Titrations

812. What is the most common reason for performing a titration?

A. To dilute an acid or base
B. To adjust the pH of an acid or base
C. To discover the concentration of an unknown solution
D. To discover the endpoint of an indicator

Use the graph below to answer questions 813-822. The graph below shows titration curves of 50 mL of the six acids, Q, R, S, T, U, and V. In each case, the titrant used was 0.1 *M* NaOH.

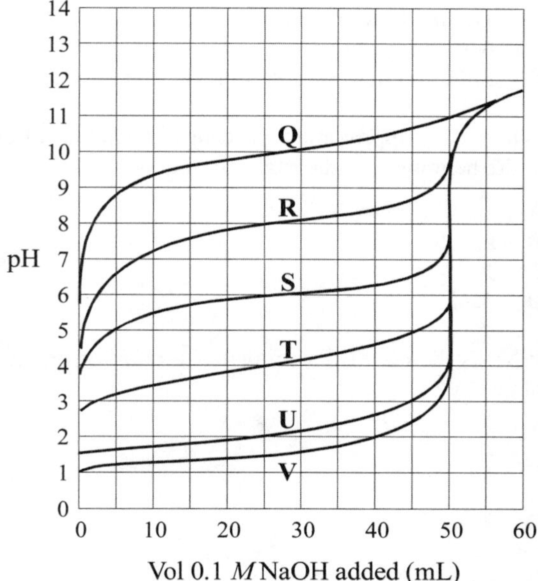

813. What was the concentration of acid S before the titration?

A. 0.05 *M*
B. 0.1 *M*
C. 0.2 *M*
D. 1 *M*

814. What is the approximate K_a of acid T?

A. 1×10^{-10}
B. 1×10^{-6}
C. 1×10^{-4}
D. 1×10^{-3}

815. What is the hydrogen ion concentration of acid U after 45 mL of 0.1 *M* NaOH have been added during the titration?

A. 0.0001 *M*
B. 0.001 *M*
C. 0.003 *M*
D. 3 *M*

816. A 0.05 *M* solution of conjugate base of acid S would have a pH of approximately:

A. 4
B. 6
C. 9
D. 12

817. The choices below give an indicator followed by the pH where a color change occurs. Which of these would be the best indicator to use for the titration of acid R?

A. Bromcresol green, 3.6-5.5
B. Thymol blue, 7.8-9.6
C. Alizarin, 8.7-11.4
D. sodium indigo sulfonate, 11.8-13.9

818. What is the approximate percent dissociation of acid S at the beginning of the titration?

A. 0.1 %
B. 2.5 %
C. 10 %
D. 25%

819. What is the pH of the equivalence point of the titration of acid V?

A. 6
B. 7
C. 8
D. 9

820. If conjugate base of acid S were titrated with acid V, which of the following represents the titration curve that would result?

A.

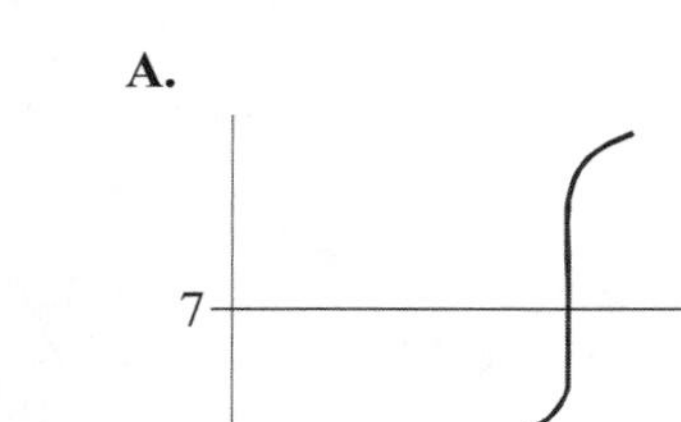

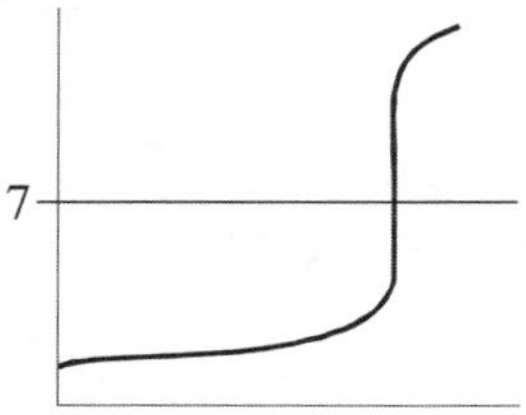

C.

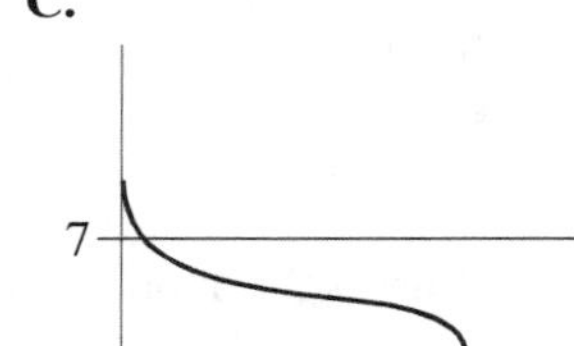

B.

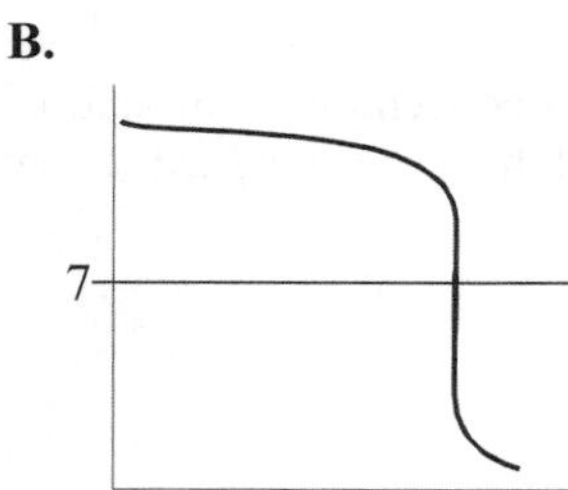

D.

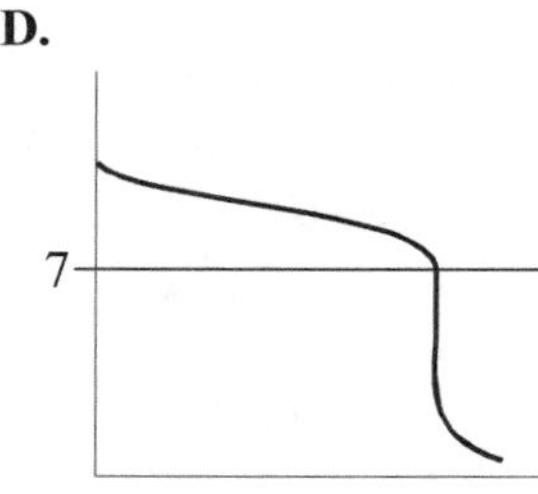

821. Acid Q is probably:

A. water
B. a weak acid
C. a weak base
D. a strong base

822. Which acid would be the most appropriate for producing a buffer solution at a pH of 4?

A. Q
B. S
C. T
D. V

---

823. 50 mL of a weak acid is titrated with 5.0 *M* NaOH. If 25 mL of NaOH is needed to reach the equivalence point, what is the molarity of the acid?

A. 2.5 *M*
B. 5.0 *M*
C. 10 *M*
D. The pK_a of the weak acid is required to answer this question

824. In a titration, 30 mL of 7 *M* NaOH is needed to reach the equivalence point with 100 mL of HF solution. What is the molarity of the HF solution?

A. 2.1 *M*
B. 3.1 *M*
C. 10 *M*
D. 23.3 *M*

825. If 5.0 *M* NaOH were titrated with 3.0 *M* HCl, which of the following would most likely resemble the titration curve?

A.

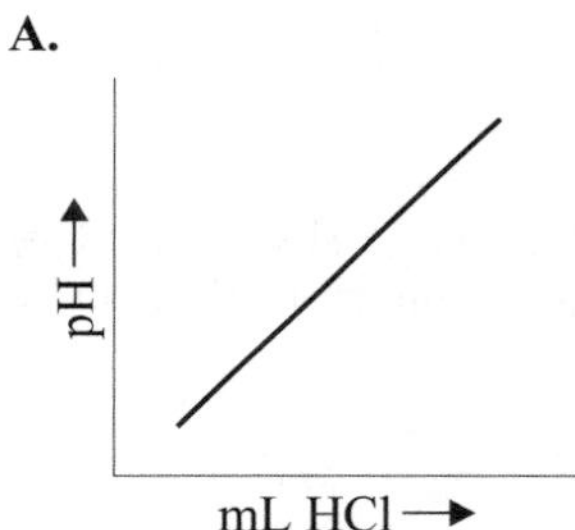

C.

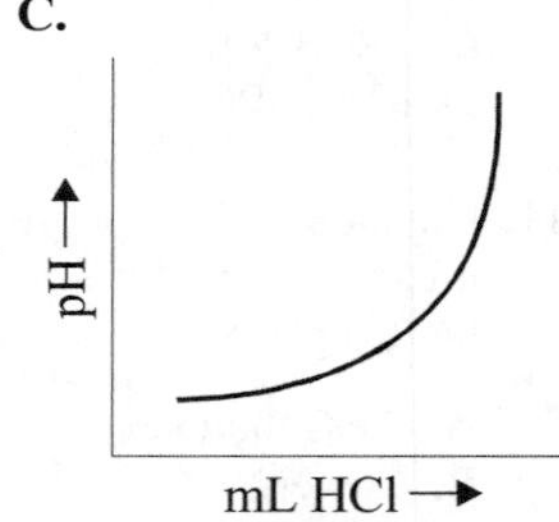

B.

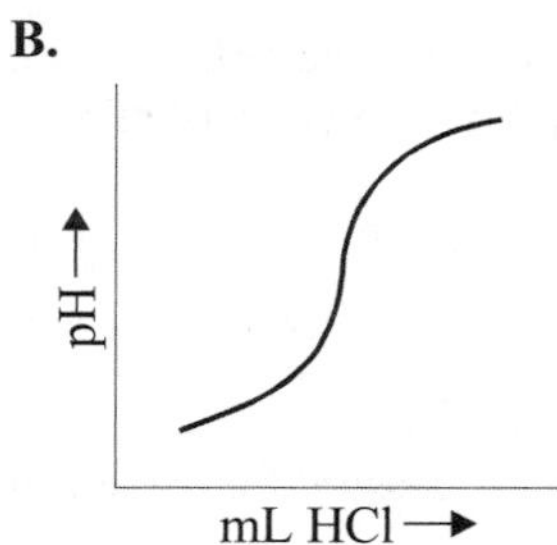

D.

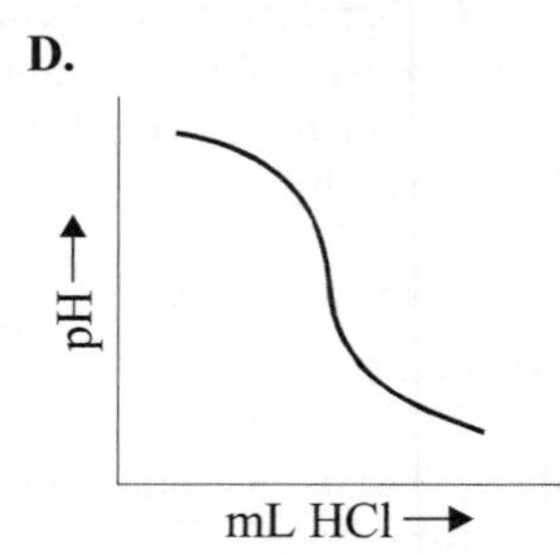

826. If 30 mL of 3 *M* acetic acid (pK_a=4.7) were added to 50 mL of 2 *M* sodium hydroxide, the resulting solution would have a pH:

A. less than two.
B. between two and seven.
C. equal to seven.
D. greater than seven.

827. If 30 mL of 3 *M* acetic acid (pK_a=4.7) were added to 45 mL of 2 *M* sodium hydroxide, the resulting solution would have a pH:

A. less than two.
B. between two and seven.
C. equal to seven.
D. greater than seven.

828. If 30 mL of 3 *M* acetic acid (pK_a=4.7) is added to 40 mL of 2 *M* sodium hydroxide, the resulting solution has a pH:

A. less than two.
B. between two and seven.
C. equal to seven.
D. greater than seven.

829. 30.0 mL of sodium cyanide is titrated with 21.0 mL of 7.0 *M* HCl. What was the concentration of the original sodium cyanide solution?

A. 4.9 *M*
B. 5.1 *M*
C. 7.0 *M*
D. 10.0 *M*

830. If 30 mL of 1.0 *M* NaOH are added to 30 mL of 5.0 *M* acetic acid (pK_a = 4.74), what are the major species present in the resulting solution?

A. Water, acetic acid, hydroxide, and sodium ions
B. Water, acetate, and sodium ions
C. Water and sodium ions
D. Water, acetic acid, acetate, and sodium ions

More Titrations and Buffered Solutions

831. Which of the following pairs would *not* make a good buffer if placed in the same aqueous solution?

A. HCl and KCl
B. H_2S and NaHS
C. NH_4Cl and NH_3
D. $KHCO_3$ and K_2CO_3

832. If an experimenter mixed an equal number of moles of hydrochloric acid (a strong acid) with ammonia (a weak base), the result would be best described by which of the following?

A. Since hydrochloric acid is a strong acid, the reaction goes nearly to completion, resulting in a solution with pH 7.
B. Since hydrochloric acid is a strong acid, the reaction goes nearly to completion, resulting in a solution with pH less than 7.
C. Since ammonia is a weak base, the reaction should be represented as an equilibrium. The unreacted hydrochloric acid then gives a solution with pH less than 7.
D. Since ammonia is a weak base, the reaction should be represented as an equilibrium. The hydrochloric acid, however, dissociates completely. The unreacted ammonia then gives a solution with pH greater than 7.

833. A solution is composed of equal concentrations of aqueous ammonia (pK_b=4.74) and ammonium chloride (pK_a=9.26). What would be the pH of the resulting solution?

A. 4.74
B. 7.00
C. 9.26
D. The pH of the solution cannot be determined from the information given

834. An aqueous solution is composed of 2 *M* ammonia (pK_b = 4.74) and 0.2 *M* ammonium chloride (pK_a = 9.26). What is the pH of the solution?

A. 4.74
B. 8.26
C. 9.26
D. 10.26

835. An aqueous solution is created from equal amounts of 1 *M* HCl (a strong acid) and 0.01 *M* sodium chloride. What is the approximate pH of the solution?

A. 0
B. 1
C. 2
D. The pH of the solution cannot be estimated without the volumes of solution.

836. Which of the following would be an equation that could be used to find the pH of a buffer if the pK_b of the base and the concentrations of base and acid are known?

A. $pH = pK_b + \log([base]/[acid])$
B. $pH = pK_b + \log([acid]/[base])$
C. $pH = pK_b - \log([base]/[acid])$
D. $pH = 14 - pK_b + \log([base]/[acid])$

837. A buffer is made by combining 30 mL of a 2.0 *M* acidic solution with 42.0 grams of its salt. The resulting solution has a pH of 7.5. If the solution is then diluted to a total volume of 300 mL, what will be the new value of the pH?

A. 6.5
B. 7.5
C. 8.5
D. Cannot be determined from the information given

838. Solution A contains a buffer formed of 2.0 *M* acetic acid and 0.8 *M* potassium acetate. Solution B contains a buffer formed of 0.2 *M* acetic acid and 0.08 *M* potassium acetate. How do the solutions compare?

A. The pH and buffer capacity of both solutions are the same
B. The pH of solution A is lower than that of solution B, but the buffer capacity is the same
C. The pH of solution A is the same as that of solution B, but solution A has a greater buffer capacity
D. The pH of solution A is lower than that of solution B, and solution A has a greater buffer capacity than solution B

839. An unknown weak acid is titrated with sodium hydroxide. Which of the following can be determined from a titration curve?

I. The initial concentration of the acid
II. The pK_a of the acid
III. The molecular weight of the acid

A. I only
B. I and II only
C. II and III only
D. I, II, and III

840. A monoprotic weak acid is titrated with a strong base. At the first equivalence point, the pH will be:

A. less than 7
B. 7
C. more than 7
D. A choice cannot be made without more information.

841. A monoprotic weak acid is titrated with a strong base. In the middle of the first buffer region, the pH will be:

A. less than 7
B. 7
C. more than 7
D. A choice cannot be made without more information.

Indicators and the End Point

842. Indicators are weak acids that change color over a fairly narrow pH range. In order to be used with a titration, an indicator should change color at a pH:

A. roughly equal to the p*K*a of the acid being titrated
B. roughly equal to the pH at the equivalence point
C. of 7
D. The pH that the indicator changes color at is irrelevant.

843. Which of the following is measured *directly* in a titration?

A. The degree of dissociation of the titrant
B. The pK_a of the indicator
C. The equivalence point for the titration
D. The endpoint of the titration

844. If the equivalence point of a titration is known to be at a pH of 7. Which of the following is the most appropriate indicator?

A. alazarin yellow, $K_a = 1.6\text{x}10^{-11}$
B. phenolphthalein, $K_a = 1.0\text{x}10^{-8}$
C. bromthymol blue, $K_a = 7.9\text{x}10^{-8}$
D. methyl orange, $K_a = 3.2\text{x}10^{-4}$

Polyprotic Titrations

845. Which of the following describes the spacing of the equivalence points when a triprotic acid is titrated with a strong base?

A. As each proton is removed, the acid becomes weaker, thus each successive equivalence point will require a smaller volume of base to be added
B. As each proton is removed, the acid becomes weaker, thus each successive equivalence point will require a larger volume of base to be added
C. The ratio of base added to proton removed in a titration is one-to-one, therefore each successive equivalence point will require the same volume of base to be added as the previous one
D. The ratio of base added to proton removed in a titration is one-to-one, therefore each successive equivalence point will experience the same change in pH as the previous one

846. An unknown acid titrated with 8.0 M NaOH produced the titration curve shown below. The acid could be:

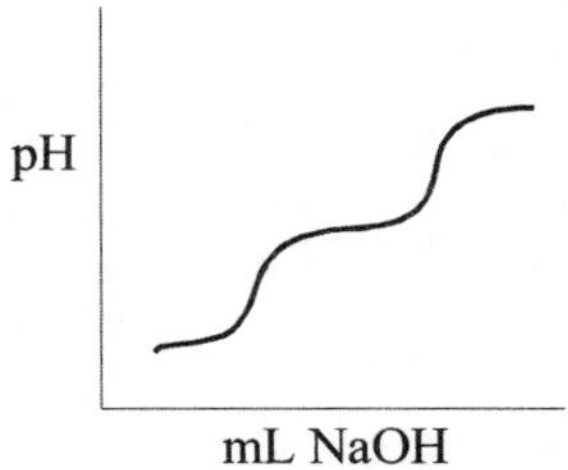

A. HF
B. HCl
C. H_2CO_3
D. H_3PO_4

Use the titration curve below to answer questions 847-853. The titration curve below shows the titration of 0.1 *M* H_2CO_3 with 0.1 *M* NaOH.

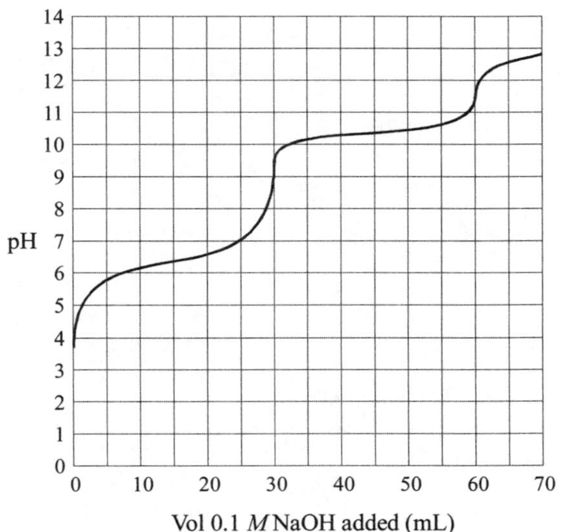

847. What is the volume of the H_2CO_3 sample being titrated?

A. 15 mL
B. 30 mL
C. 60 mL
D. 70 mL

848. If the concentration of carbonic acid were greater, how would the titration curve change?

A. The pH of the starting point would be lower and the pH of both equivalence points would be greater.
B. The pH of the starting point and both equivalence points would be lower.
C. The pH of the both half equivalence points would be greater.
D. The pH of the both half equivalence points would be lower.

849. What is the approximate pH at the second equivalence point?

A. 6.4
B. 9.5
C. 10.3
D. 11.4

850. The K_{a1} for H_2CO_3 is $4.3 \text{x} 10^{-7}$. The K_{a2} for the H_2CO_3 is $5.6 \text{x} 10^{-11}$. At approximately what pH is the concentration of HCO_3^- is equal to the concentration of CO_3^{2-}?

A. 6.4
B. 7
C. 10.3
D. 12

851. Suppose the titration is stopped after 15 mL of NaOH is added, and 10 mL of water is added to the solution. What would be the new pH?

A. 5.8
B. 6.4
C. 7
D. 9.5

852. What is the H^+ concentration at the first equivalence point?

A. 0 *M*
B. $2.9 \text{x} 10^{-10}$ *M*
C. $1 \text{x} 10^{-7}$ *M*
D. $3.2 \text{x} 10^{-8}$ *M*

853. At the second equivalence point, the respective concentrations of H_2CO_3, HCO_3^-, and CO_3^{2-} are:

A. 0 *M*, 0 *M*, and 0.033 *M*
B. 0 *M*, 0 *M*, and 0.1 *M*
C. 0 *M*, 0.05 *M*, and 0.05 *M*
D. 0.1 *M*, 0.1 *M*, and 0.1 *M*

---

854. 50.0 mL of a 3.00 *M* solution of carbonic acid is titrated with a 7.00 *M* solution of sodium hydroxide. How much of the sodium hydroxide solution is necessary to convert all of the carbonic acid to sodium carbonate?

A. 10.7 mL
B. 21.4 mL
C. 42.8 mL
D. 100.0 mL

855. A diprotic acid is titrated with a strong base. At the first equivalence point, the pH will be:

A. below 7
B. 7
C. above 7
D. Cannot be determined without more information.

856. A diprotic acid is titrated with a strong base. At the second equivalence point, the pH will be:

A. below 7
B. 7
C. above 7
D. Cannot be determined without more information.

857. All of the following are true about a diprotic acid EXCEPT:

A. It's always easier to remove the first proton than the second.
B. On a titration curve for a polyprotic acid, there is one equivalence point for each acidic proton.
C. In concentrated solutions, the second proton does not affect the pH.
D. The K_a values increase for successive acidic protons on a single polyprotic acid.

858. The K_a for HSO_4^- is 1.2×10^{-2}. When 1 mole of HSO_4^- is added to one liter of pure water:

A. the ion completely dissociates.
B. most, but not all, of the ions gain a proton.
C. most, but not all, of the ions lose a proton.
D. some of the ions lose a proton, but most of the ions neither gain nor lose a proton.

859. The K_a for HSO_4^- is 1.2×10^{-2}. At a pH of about 2, HSO_4^- concentration is about equal to the concentration of:

A. H_2SO_4
B. SO_4^{2-}
C. H^+
D. It depends upon the concentration of HSO_4^- because the percent dissociation changes with concentration.

860. The K_{a1} for H_2CO_3 is 4.3×10^{-7}. The K_{a2} for the H_2CO_3 is 5.6×10^{-11}. Which of the following depends upon the concentration of acid and conjugate base?

A. The pH of the first equivalence point
B. The pH of the first half equivalence point
C. The pH of the second half equivalence point
D. The pK_{a1}

Oxidation-Reduction

861. In an oxidation-reduction reaction:

A. compounds oxidize and reduce, while atoms are oxidized and reduced.
B. atoms oxidize and reduce, while compounds are oxidized and reduced.
C. atoms oxidize and reduce, while atoms are oxidized and reduced.
D. compounds oxidize and reduce, while compounds are oxidized and reduced.

862. $I_2O_5(s) + 5CO(g) \rightarrow I_2(s) + 5CO_2(g)$

Which of the following is true for the reaction above?

A. I_2O_5 is the oxidant and CO is oxidized.
B. I_2O_5 is the oxidant and C is oxidized.
C. Oxygen is the oxidant and CO is oxidized.
D. Oxygen is the oxidant and C is oxidized.

863. In an oxidation-reduction reaction, the atom that is oxidized always:

A. bonds to oxygen.
B. bonds to hydrogen.
C. gains electrons.
D. loses electrons.

864. When an atom is oxidized:

A. there is an actual transfer of electrons from the reductant to the atom.
B. there is an actual transfer of electrons from the atom to the reductant.
C. the oxidation state of the atom decreases.
D. the oxidation state of the atom increases.

865. Which of the following is true concerning reduction and oxidation?

A. Redox reactions are spontaneous in both directions.
B. The presence of oxygen is required for oxidation to occur.
C. In some reduction reactions no oxidation occurs.
D. An atom in the oxidizing agent is reduced.

866. Consider the following reaction:

$$CaH_2 + 2BH_3 \rightarrow 2BH_4^- + Ca^{2+}$$

What type of reaction is this?

A. Lewis acid/Lewis base
B. Brönsted-Lowry acid/base
C. Oxidation/reduction
D. Combustion

867. Consider the following reaction:

$$CaF_2 + H_2SO_4 \rightarrow CaSO_4 + 2HF$$

What type of reaction is this?

A. Brönsted-Lowry acid/base
B. Oxidation/reduction
C. Combustion
D. Decomposition

868. Consider the following steps for the Ostwald process for synthesizing nitric acid:

Step 1: $4NH_3 + 5O_2 \rightarrow 4NO + 6H_2O$
Step 2: $2NO + O_2 \rightarrow 2NO_2$
Step 3: $3NO_2 + H_2O \rightarrow 2HNO_3 + NO$

Which of the steps are redox reactions?

A. Step 3 only
B. Steps 1 and 2 only
C. Steps 1, 2, and 3
D. None of the steps is a redox reaction.

869. What is the oxidation number of chlorine in chloric acid ($HClO_3$)?

A. –1
B. 0
C. +1
D. +5

870. What is the oxidation number of iron in Fe_2S_3?

A. 0
B. +1
C. +2
D. +3

871. What is the oxidation number of phosphorus in the phosphate ion?

A. +3
B. +5
C. +8
D. +11

872. What is the oxidation number of chromium in sodium dichromate ($Na_2Cr_2O_7$)?

A. +2
B. +3.5
C. +6
D. +7

873. What is the oxidation number of xenon in potassium perxenate, K_4XeO_6?

A. +2
B. +8
C. +12
D. This compound cannot exist; it is impossible for xenon to be oxidized to this degree

874. What is the oxidation number of sodium in metallic sodium?

A. −1
B. 0
C. +1
D. The oxidation number varies.

875. What is the oxidation number of nitrogen in sodium azide, NaN_3?

A. −3
B. −1
C. −1/3
D. 0

876. Fluorine is the most electronegative element. What is the second most electronegative element?

A. nitrogen
B. chlorine
C. oxygen
D. neon

877. What is the oxidation number of oxygen in OF_2?

A. −2
B. −1/2
C. 0
D. +2

878. What is the oxidation number of oxygen in O_2F_2?

A. −1
B. 0
C. +1
D. +2

879. O_2^- is called the superoxide ion. What is the oxidation number of oxygen in O_2^-?

A. −2
B. −1
C. −1/2
D. 0

880. When an element is both oxidized and reduced in the same reaction, disproportionation is said to occur. In which of the following reactions does disproportionation occur?

A. $Hg(l) + H_2SO_4(l) \rightarrow HgSO_4(s) + SO_2(g) + 2H_2O(l)$
B. $2H_2O_2(l) \rightarrow 2H_2O(l) + O_2(g)$
C. $CaCO_3(s) + H_2O(l) + CO_2(aq) \rightarrow Ca(HCO_3)_2(aq)$
D. $SO_2(g) + H_2O(l) \rightarrow H_2SO_4(aq)$

881. Hydrogen peroxide H_2O_2 can behave as a reducing agent or an oxidizing agent as shown by the two half reactions below labeled R1 and R2. Which of the following is true?

$$R1: O_2(g) + 2H^+(aq) + 2e^- \rightarrow H_2O_2(aq) \quad E^\circ = 0.68 \text{ V}$$

$$R2: 2H^+(aq) + H_2O_2(aq) + 2e^- \rightarrow 2H_2O(l) \quad E^\circ = 1.78 \text{ V}$$

A. Reaction 1 demonstrates that hydrogen peroxide behaves as a reducing agent in basic solutions.
B. Reaction 2 demonstrates that hydrogen peroxide behaves as a reducing agent in basic solutions.
C. Reaction 1 demonstrates that hydrogen peroxide behaves as an oxidizing agent in basic solutions.
D. Reaction 2 demonstrates that hydrogen peroxide behaves as an oxidizing agent in basic solutions.

882. In the compound $KHSO_4$, which element has the highest oxidation number?

A. K
B. H
C. S
D. O

883. If a metal forms more than one oxide, the basicity decreases as the oxidation state of the metal increases. Which of the following is the strongest base?

A. CrO
B. CrO_2
C. CrO_3
D. Cr_2O_3

884. Which of the following gives the oxides of nitrogen in order of increasing nitrogen oxidation number?

A. NO, NO_2, N_2O, N_2O_3, N_2O_5
B. N_2O_5, N_2O_3, N_2O, NO_2, NO
C. NO, N_2O, NO_2, N_2O_3, N_2O_5
D. N_2O, NO, N_2O_3, NO_2, N_2O_5

885. Although oxygen often has an oxidation state of –2, it can occasionally take on other values. In which of the following choices do the oxygen atoms possess the oxidation numbers –2, –1, –1/2, and 0, in that order?

A. O, O_2, O_3, O_4
B. CaO, Na_2O, Al_2O_3, O_2
C. Na_2O, NaO, NaO_2, O_3
D. HClO, $HClO_2$, $HClO_3$, $HClO_4$

886. Consider the following redox reaction:

$$Cl_2 + Fe \rightarrow Fe^{2+} + 2Cl^-$$

Which of the following is the reducing agent in this reaction?

A. Cl_2
B. Fe
C. Fe^{2+}
D. Cl^-

887. Consider the following reaction:

$$FeI_2 + F_2 \rightarrow FeF_2 + I_2$$

Which of the following is the reducing agent in this reaction?

A. Fe
B. FeI_2
C. F_2
D. This reaction does not involve reduction

888. Consider the following reaction:

$$FeI_2 + VF_2 \rightarrow FeF_2 + VI_2$$

Which of the following is the reducing agent in this reaction?

A. Fe
B. FeI_2
C. VF_2
D. This reaction does not involve reduction

889. Consider the following reaction:

$$2NaClO_3 + SO_2 + H_2SO_4 \rightarrow 2NaHSO_4 + 2ClO_2$$

Which of the following is/are the reducing agent(s) in this reaction?

I. $NaClO_3$
II. SO_2
III. H_2SO_4

A. I only
B. II only
C. III only
D. II and III only

890. Consider the following reaction:

$$2NaClO_3 + SO_2 + H_2SO_4 \rightarrow 2NaHSO_4 + 2ClO_2$$

Which of the following is/are the oxidizing agent(s) in this reaction?

I. $NaClO_3$
II. SO_2
III. H_2SO_4

A. I only
B. II only
C. III only
D. II and III only

891. Consider the following reaction:

$$MnS + CrPO_4 \rightarrow CrS + MnPO_4$$

Which of the following is/are the reducing agent(s) in this reaction?

A. Mn
B. MnS
C. $CrPO_4$
D. This reaction does not involve reduction.

892. Water is capable of acting as a/an:

I. Brönsted-Lowry base
II. Lewis acid
III. Oxidizing agent

A. I only
B. I and II only
C. I and III only
D. I, II, and III

893. Consider the following reaction:

$$Fe^{3+} + 6H_2O \rightarrow [Fe(H_2O)_6]^{3+}$$

In this reaction, water acts as a/an:

A. Lewis acid.
B. Lewis base.
C. oxidizing agent.
D. reducing agent.

894. Consider the following reaction:

$$2Na + 2H_2O \rightarrow 2NaOH + H_2$$

In this reaction, water acts as a/an:

A. Brönsted-Lowry acid.
B. Brönsted-Lowry base.
C. oxidizing agent.
D. reducing agent.

895. Consider the following reaction:

$$NaH + H_2O \rightarrow NaOH + H_2$$

What is the role of water in this redox reaction?

A. It is an oxidizing agent only
B. It is a reducing agent only
C. It is both an oxidizing and a reducing agent
D. It is neither an oxidizing nor a reducing agent

896. Consider the following reaction:

$$3NO_2 + H_2O \rightarrow 2HNO_3 + NO$$

What is the role of water in this redox reaction?

A. It is an oxidizing agent only
B. It is a reducing agent only
C. It is both an oxidizing and a reducing agent
D. It is neither an oxidizing nor a reducing agent

897. Consider the following reaction:

$$3NO_2 + H_2O \rightarrow 2HNO_3 + NO$$

What is the role of NO_2 in this redox reaction?

A. It is an oxidizing agent only
B. It is a reducing agent only
C. It is both an oxidizing and a reducing agent
D. It is neither an oxidizing nor a reducing agent

898. Which of the following is necessary to transform nitric acid (HNO_3) into nitrous acid (HNO_2)?

A. An acid
B. A base
C. An oxidizing agent
D. A reducing agent

899. Which of the following is necessary to transform the nitrite ion (NO_2^-) into nitrous acid (HNO_2)?

A. An acid
B. A base
C. An oxidizing agent
D. A reducing agent

900. Which of the following is necessary to transform nitrogen gas into ammonia gas?

A. An acid
B. A base
C. An oxidizing agent
D. A reducing agent

901. Which of the following is NOT a possible oxidation number for carbon?

A. –4
B. 0
C. +2
D. +6

Oxidation-Reduction Titrations

902. In a redox titration a titration curve can be generated comparing voltage and volume of titrant. In order to generate a titration curve for a redox titration, which of the following must be true?

A. The reducing agent must be the titrant.
B. The oxidizing agent must be the titrant.
C. A standard solution must be provided against which the voltage can be compared.
D. An indicator must be used.

903. In a redox titration, if the voltage in the solution rises:

A. The reducing agent must be the titrant.
B. The oxidizing agent must be the titrant.
C. The titrant may be either the reducing agent or the oxidizing agent.
D. The voltage of the standard solution must be decreasing.

Use the following information to answer questions 904-912. The amount of iron in iron ore can be found by dissolving the iron ore in acid solution to produce aqueous Fe^{2+} ions and then titrating with potassium permanganate ($KMnO_4$). Potassium permanganate can act as its own indicator. The MnO_4^- ion is purple, while the Mn^{2+} ion is almost colorless.

904. A student dissolves a sample of iron ore in acid. Which of the following would be the correct titration procedure to discover the concentration of Fe^{3+} in solution?

A. A known concentration of MnO_4^- solution should be dripped into the iron solution until the solution turns purple.
B. A known concentration of MnO_4^- solution should be dripped into the iron solution until the solution turns colorless.
C. A known concentration of Mn^{2+} solution should be dripped into the iron solution until the solution turns purple.
D. A known concentration of Mn^{2+} solution should be dripped into the iron solution until the solution turns colorless.

905. Which of the following half reactions involving Fe^{2+} most likely takes place in the solution during the titration?

A. $Fe^{2+}(aq) + 2e^- \rightarrow Fe(s)$
B. $Fe^{3+}(aq) + e^- \rightarrow Fe^{2+}(aq)$
C. $Fe^{2+}(aq) \rightarrow Fe^{3+}(aq) + e^-$
D. $Fe^{2+}(aq) \rightarrow Fe^{4+}(aq) + 2e^-$

906. Which of the following half reactions involving potassium permanganate most likely takes place in the solution during the titration?

A. $8H^+(aq) + MnO_4^-(aq) + 5e^- \rightarrow Mn^{2+}(aq) + H_2O(l)$
B. $Mn^{7+}(aq) + 5e^- \rightarrow Mn^{2+}(aq)$
C. $MnO_4^-(aq) \rightarrow Mn^{2+}(aq) + 3e^-$
D. $Mn^{2+}(aq) + 3e^- \rightarrow MnO_4^-(aq)$

907. The balanced redox reaction is:

A. $4H_2O(l) + MnO_4^-(aq) + 5Fe^{2+}(aq) \rightarrow 5Fe^{3+}(aq) + Mn^{2+}(aq) + 8H^+(aq)$
B. $8OH^-(aq) + MnO_4^-(aq) + Fe^{2+}(aq) \rightarrow Fe^{3+}(aq) + Mn^{2+}(aq) + 4H_2O(l)$
C. $8H^+(aq) + MnO_4^-(aq) + 5Fe^{2+}(aq) \rightarrow 5Fe^{3+}(aq) + Mn^{2+}(aq) + 4H_2O(l)$
D. $MnO_4^-(aq) + 5Fe^{2+}(aq) \rightarrow 5Fe^{3+}(aq) + Mn^{2+}(aq) + 2O_2(g)$

908. The balanced reaction for the titration involves the transfer of how many electrons?

A. 1
B. 2
C. 3
D. 5

909. How many iron atoms are required to reduce one permanganate ion?

A. 1
B. 2
C. 3
D. 5

910. If 20 mL of 0.01 *M* potassium permanganate are required to reach the equivalence point in the titration, how many moles of Fe^{2+} were in the original solution?

A. 0.0002
B. 0.001
C. 0.002
D. 0.004

911. Suppose the titration begins with 40 mL of Fe^{2+} solution. If 20 mL of 0.01 *M* potassium permanganate are required to reach the equivalence point in the titration, what was the original concentration of Fe^{2+} in solution?

A. 0.010
B. 0.025
C. 0.030
D. 0.040

912. Suppose that a 0.56 g piece of iron ore is dissolved in 40 mL of acid solution. If 20 mL of 0.01 *M* potassium permanganate are required to reach the equivalence point in the titration, what was the mass percent of iron in the iron ore?

A. 5 %
B. 10 %
C. 50 %
D. 100 %

---

913. Which of the following is the correctly balanced reaction between silver and nitric acid?

A. $Ag + H_2SO_4 \rightarrow Ag^+ + SO_3^- + H_2O$
B. $Ag + HNO_3 + 3H_3O^+ \rightarrow Ag^+ + NO + 5H_2O$
C. $Ag + H_3N + H_2O \rightarrow Ag^+ + NH_4OH$
D. $3Ag + NO_3^- + 4H_3O^+ \rightarrow 3Ag^+ + NO + 6H_2O$

914. Consider the following half-reactions:

$$H_2C_2O_4 \rightarrow 2CO_2 + 2H^+ + 2e^-$$

$$HAsO_2 + 3H^+ + 3e^- \rightarrow As + 2H_2O$$

Which of the following is the balanced equation for the reaction between $H_2C_2O_4$ and $HAsO_2$?

A. $H_2C_2O_4 + HAsO_2 \rightarrow 2CO_2 + As + H_2O$
B. $H_2C_2O_4 + HAsO_2 + H^+ \rightarrow 2CO_2 + As + 2H_2O$
C. $H_2C_2O_4 + As + 2H_2O \rightarrow 2CO_2 + 5H^+ + HAsO_2$
D. $3H_2C_2O_4 + 2HAsO_2 \rightarrow 6CO_2 + 2As + 4H_2O$

915. Suppose a galvanic cell operates using the following half-reactions:

Anode:

$$Zn(s) + 4NH_3(g) \rightarrow [Zn(NH_3)_4]^{2+}(aq) + 2e^-$$

Cathode:

$$MnO_2(s) + NH_4^+(aq) + e^- \rightarrow MnO(OH)(s) + NH_3(g)$$

If 3 moles of Zn are oxidized, how many moles of manganese (IV) oxide are reduced?

A. 1.5
B. 3
C. 4.5
D. 6

916. Suppose an electrolytic cell utilizes the following half-reactions:

Anode:

$$Mg \rightarrow Mg^{2+} + 2e^-$$

Cathode:

$$Cu^+ + e^- \rightarrow Cu$$

How many moles of magnesium need to be oxidized in order to reduce three moles of copper ions?

A. 2/3
B. 3/2
C. 3
D. 6

917. If a current of 3 amperes is used to reduce Cu^{2+} to Cu, how many seconds will it take for 5 grams of copper to be reduced? (The charge on one mole of electrons is equal to 96,485 C.)

A. $\frac{(5)(3)(96485)}{(63.5)}$

B. $\frac{(5)(96485)}{(3)(63.5)}$

C. $\frac{(2)(5)(96485)}{(3)(63.5)}$

D. $\frac{(5)(96485)}{(2)(3)(63.5)}$

918. Suppose an electrolytic cell utilizes the following cathode reaction to silver-plate a piece of jewelry:

$$Ag^+ + e^- \rightarrow Ag \qquad E° = +0.80 \text{ V}$$

If 0.108 grams of silver is to be plated using a 2.0 A current, for how long should the cell be run? (One Faraday = 96,485 C/mol e^-.)

A. 48 seconds
B. 108 seconds
C. 3 minutes
D. 17 minutes

Potentials

919. Suppose a student said, "The standard potential of a half-reaction can be used to calculate the equilibrium constant of the half-reaction." Which of the following statements most accurately characterizes the student's comment?

A. The statement is a consequence of the Nernst equation, and is therefore true.
B. The statement is a consequence of the definition of standard potential, and is therefore true.
C. The equilibrium constant cannot be calculated unless the conditions under which the standard potential was measured is known
D. It is meaningless to refer to the equilibrium constant of a half-reaction.

920. Which of the following is true concerning oxidation and reduction half-reactions?

A. A reduction half reaction can occur by itself if the reduction half-reaction potential is applied across the reaction.
B. A reduction half reaction can occur by itself if the negative of the reduction half-reaction potential is applied across the reaction.
C. Oxidation and reduction must take place simultaneously.
D. Reduction can only occur by itself in the presence of a strong reducing agent.

Questions 921-923 refer to the following table:

Half-reaction	E°
$2F^- \rightarrow F_2 + 2e^-$	–2.87 V
$2Cl^- \rightarrow Cl_2 + 2e^-$	–1.36 V
$Hg^{2+} + 2e^- \rightarrow Hg$	+0.85 V
$Au^{3+} + 3e^- \rightarrow Au$	+1.52 V

921. Based on the table, which of the following is the strongest reducing agent?

A. F^-
B. Cl^-
C. Hg^{2+}
D. Au^{3+}

922. What is the oxidation potential of solid gold?

A. –0.51 V
B. –1.52 V
C. +1.52 V
D. +4.56 V

923. What is the oxidation potential of a fluoride ion?

A. –1.44 V
B. –2.87 V
C. +2.87 V
D. +5.74 V

---

924. Which of the following elements can NOT be oxidized by ozone?

A. Ni
B. Fe
C. Cu
D. Au

925. Substances that are strong oxidizing agents are reduced to substances that are:

A. weak reducing agents.
B. strong reducing agents.
C. weak oxidizing agents.
D. strong oxidizing agents.

926. Magnesium disks are often attached to the (iron) hulls of ships to prevent them from rusting. Why does this work?

A. Magnesium is a better oxidizing agent than iron, thus the magnesium is oxidized preferentially.
B. Magnesium is a better oxidizing agent than iron, thus the magnesium is reduced preferentially.
C. Magnesium is a better reducing agent than iron, thus the magnesium is oxidized preferentially.
D. Magnesium is a better reducing agent than iron, thus the magnesium is reduced preferentially.

927. Suppose the amount of material undergoing an oxidation-reduction reaction doubles, but all other conditions remain the same. Which of the following also doubles?

I. Free energy change
II. Total charge flow
III. Potential

A. I only
B. II only
C. I and II only
D. II and III only

928. Which of the following is/are *not* generally found in nature?

I. $Na(s)$
II. $F_2(g)$
III. $UO_2(s)$

A. III only
B. I and II only
C. I and III only
D. II and III only

Use the table below to answer questions 929-934. The table below is an activity series of metals in aqueous solution. The half reactions are listed highest potential to lowest potential from top to bottom.

METAL	OXIDATION REACTION
Lithium	$Li \rightarrow Li^+ + e^-$
Potassium	$K \rightarrow K^+ + e^-$
Magnesium	$Mg \rightarrow Mg^{2+} + 2e^-$
Aluminum	$Al \rightarrow Al^{3+} + 3e^-$
Zinc	$Zn \rightarrow Zn^{2+} + 2e^-$
Hydrogen	$H_2 \rightarrow 2H^+ + 2e^-$
Copper	$Cu \rightarrow Cu^{2+} + 2e^-$
Silver	$Ag \rightarrow Ag^+ + e^-$

929. Which of the following metals will not form hydrogen gas when placed in acid solution?

A. Li
B. Al
C. Zn
D. Cu

930. Which of the following is the strongest oxidizing agent?

A. Li
B. 1 *M* Li^+
C. Ag
D. 1 *M* Ag^+

931. Which of the following is the strongest reducing agent?

A. Li
B. 1 *M* Li^+
C. Ag
D. 1 *M* Ag^+

932.

$$X(s) + Zn(NO_3)_2(aq) \rightarrow X(NO_3)_2(aq) + Zn(s)$$

Which of the following metals could be ***X*** in the reaction above?

A. Al
B. Mg
C. Cu
D. Ag

933. A copper wire is placed in a solution of $Ag(NO_3)$. Is there a reaction?

A. No, because silver ions cannot oxidize copper.
B. No, because silver ions cannot reduce copper.
C. Yes, because silver ions can reduce copper.
D. Yes, because silver ions can oxidize copper.

934. Nitric acid dissolves silver to produce NO gas instead of hydrogen gas. A copper wire is placed in a nitric acid solution (HNO_3). Is a gas produced?

A. No, because H^+ ions cannot oxidize copper.
B. No, because silver is more easily oxidized than copper.
C. Yes, nitrate ions oxidize copper producing NO gas.
D. Yes, nitrate ions oxidize copper producing H_2 gas.

Balancing Redox Reactions

935. Put the following steps in correct order for balancing a redox reaction:

I. Divide the reaction into half reactions.
II. Add the half reactions and simplify.
III. Balance the electrons.
IV. Balance each half reaction.

A. I, II, III, then IV
B. II, IV, I, then III
C. I, IV, III, then II
D. III, II, IV, then I

936. Put the following steps in correct order for balancing a half reaction:

I. Balance the H atoms by adding H^+.
II. Balance the elements other than H and O.
III. Balance the charge by adding e^-.
IV. Balance the O atoms by adding H_2O.

A. I, II, III, then IV
B. II, IV, I, then III
C. II, I, IV, then III
D. III, II, IV, then I

Answer questions 937-943 using the following unbalanced redox reaction occurring in basic solution.

$$CN^-(aq) + MnO_4^-(aq) \rightarrow CNO^-(aq) + MnO_2(s)$$

937. What is the unbalanced reduction half reaction?

A. $N^-(aq) \rightarrow N^{-3}(aq)$
B. $CN^-(aq) \rightarrow CNO^-(aq)$
C. $O(l) \rightarrow O^{2-}(aq)$
D. $MnO_4^-(aq) \rightarrow MnO_2(s)$

938. What is the unbalanced oxidation half reaction?

A. $N^-(aq) \rightarrow N^{-3}(aq)$
B. $CN^-(aq) \rightarrow CNO^-(aq)$
C. $O(l) \rightarrow O^{2-}(aq)$
D. $MnO_4^-(aq) \rightarrow MnO_2(s)$

939. What is the balanced reduction half reaction as it would take place in acid solution?

A. $e^- + 2H^+(aq) + MnO_4^-(aq) \rightarrow MnO_2(s) + H_2O(l)$
B. $3e^- + 4H^+(aq) + MnO_4^-(aq) \rightarrow MnO_2(s) + 2H_2O(l)$
C. $e^- + H_2O(l) + MnO_4^-(aq) \rightarrow 2H^+(aq) + MnO_2(s)$
D. $5e^- + 6H^+(aq) + MnO_4^-(aq) \rightarrow MnO_2(s) + 3H_2O(l)$

940. What is the balanced oxidation half reaction as it would take place in acid solution?

A. $e^- + 2H^+(aq) + CN^-(aq) \rightarrow CNO^-(aq) + H_2O(l)$
B. $CN^-(aq) + 2H_2O(l) \rightarrow CNO^-(aq) + 4H^+(aq) + 3e^-$
C. $CN^-(aq) + H_2O(l) \rightarrow CNO^-(aq) + H^+(aq) + e^-$
D. $CN^-(aq) + H_2O(l) \rightarrow CNO^-(aq) + 2H^+(aq) + 2e^-$

941. The next step toward balancing the equation is to add OH^- ions because the reaction takes place in basic solution. Considering only the oxidation half reaction, how should OH^- ions be added to the reaction?

A. two to the left side of the reaction
B. two to the right side of the reaction
C. two to both sides of the reaction
D. four to both sides of the reaction

942. Considering only the reduction half reaction, how should OH^- ions be added to the reaction?

A. two to the left side of the reaction
B. two to the right side of the reaction
C. two to both sides of the reaction
D. four to both sides of the reaction

943. What is the simplified balanced reaction?

A. $6OH^-(aq) + 3CN^-(aq) + 4H_2O(l) + 2MnO_4^-(aq) \rightarrow 3CNO^-(aq) + 3H_2O(l) + 2MnO_2(s) + 8OH^-(aq)$

B. $3e^- + 2OH^-(aq) + CN^-(aq) + 2H_2O(l) + MnO_4^-(aq) \rightarrow CNO^-(aq) + H_2O(l) + MnO_2(s) + 8OH^-(aq) + 2e^-$

C. $3CN^-(aq) + H_2O(l) + 2MnO_4^-(aq) \rightarrow 3CNO^-(aq) + 2MnO_2(s) + 2OH^-(aq)$

D. $3CN^-(aq) + 2MnO_4^-(aq) \rightarrow 3CNO^-(aq) + 2MnO_2(s)$

944.

$$\boldsymbol{S}H^+(aq) + \boldsymbol{T}H_2O(l) + \boldsymbol{U}NO_3^-(aq) + \boldsymbol{V}Cu(s) \rightarrow \boldsymbol{W}NO_2(g) + \boldsymbol{X}Cu^{2+}(aq) + \boldsymbol{Y}H_2O(l) + \boldsymbol{Z}H^+(aq)$$

The redox reaction above takes place in acidic solution. What are the coefficients ***S***, ***T***, ***U***, ***V***, ***W***, ***X***, ***Y***, and ***Z*** in the balanced reaction?

A. *S*=0, *T*=0, *U*=1, *V*=2, *W*=1, *X*=2, *Y*=2, *Z*=2
B. *S*=4, *T*=0, *U*=2, *V*=1, *W*=2, *X*=1, *Y*=2, *Z*=0
C. *S*=4, *T*=0, *U*=1, *V*=2, *W*=1, *X*=2, *Y*=2, *Z*=0
D. *S*=4, *T*=2, *U*=2, *V*=1, *W*=2, *X*=1, *Y*=0, *Z*=0

945.

$$\boldsymbol{R}H^+(aq) + \boldsymbol{S}H_2O(l) + \boldsymbol{T}NaBiO_3(s) + \boldsymbol{U}Mn^{2+}(aq) \rightarrow \boldsymbol{V}Na^+(aq) + \boldsymbol{W}Bi^{3+}(aq) + \boldsymbol{X}MnO_4^-(aq) + \boldsymbol{Y}H_2O(l) + \boldsymbol{Z}H^+(aq)$$

The redox reaction above takes place in acidic solution. What are the coefficients *V*, ***W***, and ***X*** in the balanced reaction?

A. *V*=2, *W*=3, *X*=2
B. *V*=3, *W*=3, *X*=1
C. *V*=5, *W*=5, *X*=2
D. *V*=4, *W*=5, *X*=1

946.

$$\boldsymbol{S}OH^-(aq) + \boldsymbol{T}H_2O(l) + \boldsymbol{U}NO_2^-(aq) + \boldsymbol{V}Al(s) \rightarrow \boldsymbol{W}NH_3(aq) + \boldsymbol{X}Al(OH)_4^-(aq) + \boldsymbol{Y}H_2O(l) + \boldsymbol{Z}OH^-(aq)$$

The redox reaction above takes place in basic solution. What are the coefficients ***W*** and ***X*** in the balanced reaction?

A. *W*=1, *X*=2
B. *W*=2, *X*=1
C. *W*=2, *X*=2
D. *W*=2, *X*=3

947.

$$\boldsymbol{S}OH^-(aq) + \boldsymbol{T}H_2O(l) + \boldsymbol{U}Cr(OH)_3(s) + \boldsymbol{V}ClO^-(aq) \rightarrow \boldsymbol{W}CrO_4^{2-}(aq) + \boldsymbol{X}Cl_2(g) + \boldsymbol{Y}H_2O(l) + \boldsymbol{Z}OH^-(aq)$$

The redox reaction above takes place in basic solution. What are the coefficients ***W*** and ***X*** in the balanced reaction?

A. *W*=1, *X*=2
B. *W*=2, *X*=1
C. *W*=2, *X*=2
D. *W*=2, *X*=3

Galvanic Cell

948. The purpose of a galvanic cell is to:

A. metal plate.
B. purify solids.
C. transduce chemical energy to electrical energy.
D. allow for oxidation without reduction.

949. An electron has a higher potential energy when it is at which electrode in a galvanic cell?

A. The anode because electrons flow toward the anode.
B. The anode because electrons flow away from the anode.
C. The cathode because electrons flow toward the cathode.
D. The cathode because electrons flow away from the cathode.

950. In a galvanic cell anions and cations move through the salt bridge, while electrons move through the load. In a galvanic cell:

A. *anions* and *electrons* move toward the *anode.*
B. *anions* and *electrons* move toward the *cathode.*
C. *anions* move toward the *cathode* and *electrons* move toward the *anode.*
D. *anions* move toward the *anode* and *electrons* move toward the *cathode.*

951. In a first experiment, a solid zinc strip is placed in copper sulfate solution. The strip dissolves and a solid copper forms in its place. In a second experiment, a beaker of one molar copper sulfate solution is connected to a beaker of one molar zinc sulfate solution by a salt bridge. A zinc strip is placed in the zinc sulfate solution and a copper strip is placed in the copper sulfate solution. What happens when the strips are connected by a copper wire?

A. Electrons flow from the *cathode* in the *zinc* sulfate solution to the *anode* in the *copper* sulfate solution.
B. Electrons flow from the *anode* in the *zinc* sulfate solution to the *cathode* in the *copper* sulfate solution.
C. Electrons flow from the *cathode* in the *copper* sulfate solution to the *anode* in the *zinc* sulfate solution.
D. Electrons flow from the *anode* in the *copper* sulfate solution to the *cathode* in the *zinc* sulfate solution.

952. The emf of a galvanic cell depends upon all the following EXCEPT:

A. the temperature of the solutions in the half cells.
B. the concentrations of the solutions in the half cells.
C. the reactions in the solutions in the half cells.
D. the length of the wire connecting the half cells.

Questions 953-956 depend on the following table:

Half-reaction	E°, V
$Ag^+ + e^- \rightarrow Ag$	+0.80
$Cd^{2+} + 2e^- \rightarrow Cd$	−0.40
$2F^- \rightarrow F_2 + 2e^-$	−2.87
$2I^- \rightarrow I_2 + 2e^-$	−0.54
$Na^+ + e^- \rightarrow Na$	−2.71
$Se^{2-} \rightarrow Se + 2e^-$	+0.67

953. Based only on the half-reactions in the table, which of the following is the strongest oxidizing agent?

A. Cd^{2+}
B. F^-
C. I^-
D. Na^+

954. Based only on the half-reactions in the table, which of the following is the strongest reducing agent?

A. Cd^{2+}
B. F^-
C. I^-
D. Na^+

955. What would be the standard cell potential of a galvanic cell in which sodium and iodine electrodes are placed in a 1 *M* sodium iodide solution?

A. 1.00 V
B. 2.17 V
C. 3.25 V
D. 5.96 V

956. In a galvanic cell in which sodium and iodine electrodes are placed in a 1 *M* sodium iodide solution, which reaction would most likely take place at the cathode?

A. $2I^- \rightarrow I_2 + 2e^-$
B. $Na^+ + e^- \rightarrow Na$
C. $I_2 + 2e^- \rightarrow 2I^-$
D. $Na \rightarrow Na^+ + e^-$

---

957. Consider an electrochemical cell that utilizes lithium and zinc:

$Li^+ + e^- \rightarrow Li$ E° = –3.05 V

$Zn^{2+} + 2e^- \rightarrow Zn$ E° = –0.76 V

What is the standard voltage for this cell?

A. 1.90 V
B. 2.29 V
C. 3.81 V
D. 5.34 V

958. In an ordinary lead automobile battery, lead anode and lead(IV)oxide cathodes are placed directly in a solution of sulfuric acid. The half-reactions are:

$$Pb(s) + SO_4^{2-} \rightarrow PbSO_4(s) + 2e^-$$

$$PbO_2(s) + H_2SO_4(aq) + 2H^+ + 2e^- \rightarrow PbSO_4(s) + 2H_2O(l)$$

Why is a salt bridge unnecessary?

A. Lead sulfate is insoluble in water, so it does not contribute additional ions to solution.
B. The acidic conditions would dissolve a salt bridge.
C. Lead and lead (IV) oxide are insoluble in water, so they do not have to be kept in separate half-cells to prevent them from reacting.
D. The electrons generated by the battery are consumed by the electrical equipment of the car.

Free Energy and Chemical Energy

959. Which expression gives the relationship between free energy and emf?

A. $\Delta G = nFE$
B. $\Delta G = -nFE$
C. $\Delta G = -nFE^o$
D. $\Delta G^o = -nFE$

960. Which of the following reactions might NOT be spontaneous?

A. A reaction where ΔG is negative.
B. A reaction where E is positive.
C. A reaction where K is greater than one.
D. A reaction in a galvanic cell.

961. A spontaneous redox reaction is indicated by:

A. a positive emf.
B. a negative emf.
C. a positive enthalpy.
D. a negative enthalpy.

962. Which of the following is the maximum amount of non pressure volume work available from a galvanic cell?

A. negative ΔG
B. negative ΔG^o
C. E
D. E^o

963. The equation below gives the relationship between ΔG and ΔG^o.

$$\Delta G = \Delta G^o + RT\ln(Q)$$

Which of the following follows from this relationship?

A. ΔG is greater than ΔG^o.
B. At equilibrium, ΔG is equal to ΔG^o.
C. ΔG is equal to $-RT\ln(K)$.
D. ΔG^o is equal to $-RT\ln(K)$.

964. Which of the following represents ΔG of a galvanic cell when all concentrations in the half cell solutions are 1.4 *M*? (Note: n = number of electrons transferred in a balanced redox reaction; F = 96,500 C/mol; E = the emf of the cell.)

A. nFE
B. $-nFE$
C. nFE^o
D. $-nFE^o$

Us the table below to answer questions 965-967.

Half-reaction	E^o, V
$Cu^{2+} + 2e^- \rightarrow Cu$	0.34
$Zn^{2+} + 2e^- \rightarrow Zn$	−0.76

965. A galvanic cell is made using 2 *M* Zn^{2+} and 1 *M* Cu^{2+} solutions. The emf of the cell is:

A. less than 1.1 V
B. 1.1 V
C. greater than 1.1 V
D. The answer cannot be determined from the information given.

966. A galvanic cell is made using 1 *M* Zn^{2+} and 2 *M* Cu^{2+} solutions. The emf of the cell is:

A. less than 1.1 V
B. 1.1 V
C. greater than 1.1 V
D. The answer cannot be determined from the information given.

967. A galvanic cell is made using 2 *M* Zn^{2+} and 2 *M* Cu^{2+} solutions. The emf of the cell is:

A. less than 1.1 V
B. 1.1 V
C. greater than 1.1 V
D. The answer cannot be determined from the information given.

968. A galvanic cell has a potential difference V at standard conditions. Which of the following graphs represents the voltage vs. time as current flows in the cell?

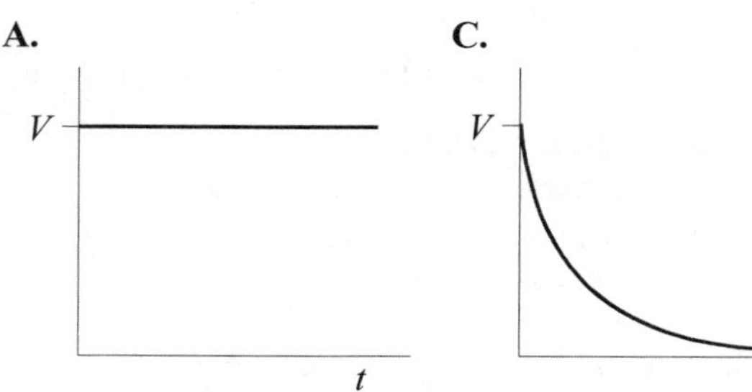

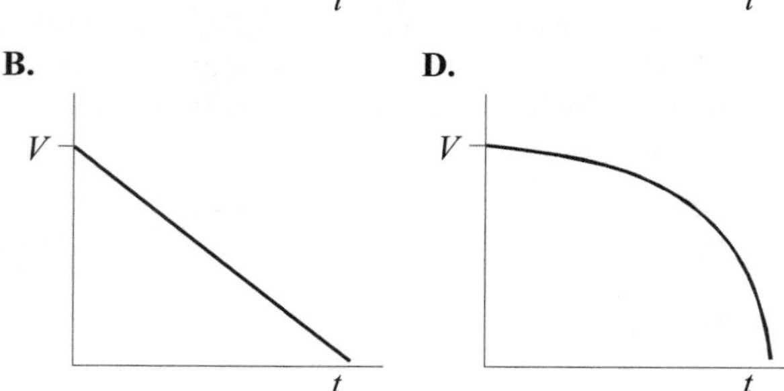

969. Consider a cell making use of the following half-reactions:

Anode:

$$Mg \rightarrow Mg^{2+} + 2e^{-}$$

Cathode:

$$MnO_2 + 4H^{+} + 2e^{-} \rightarrow Mn^{2+} + 2H_2O$$

Under standard conditions, this cell yields a potential of 3.6 V. If the pH were then increased to 7 and the concentrations of the other ions maintained at 1 *M*, the cell potential would most likely:

A. Decrease
B. Increase
C. Stay the same
D. Asymptotically approach 1 V

970. Many electrolytic cells involve reactions with positive free energies of reaction. How is this possible?

A. Chemical thermodynamics does not apply to electrical systems.
B. The entropy of free electrons is not accounted for in reaction free energies.
C. The nonspontaneous reactions are driven by spontaneous reactions.
D. Since electrons are negatively charged, the flow of free energy is reversed.

971. The standard potential of a galvanic cell is +2.03 V. Which of the following must be true of the cell?

I. $K > 1$
II. $\Delta S° > 0$
III. $\Delta G° < 0$

A. II only
B. II and III only
C. I and III only
D. I, II, and III

972. A galvanic cell with a standard cell potential of –0.02 V is designed so as to operate spontaneously when first used. Which of the following statements is/are true for this cell when first used?

I. $K > 1$
II. $Q < K$
III. $\Delta G° < 0$

A. II only
B. III only
C. I and III only
D. I, II, and III

973. Which of the following statements is/are true for a galvanic cell at equilibrium at 298 K?

I. $Q = K$
II. $E = E°$
III. $\Delta G° = \Delta G$

A. I only
B. III only
C. II and III only
D. I, II, and III

974. The standard potential of a reaction is +0.38 V. Which of the following can be concluded about this reaction?

A. Under standard conditions, the free energy of reaction is positive
B. At equilibrium, the free energy of the reaction is positive
C. Under standard conditions, the free energy of reaction is negative
D. At equilibrium, the free energy of the reaction is negative

975. Which of the following will ALWAYS increase the emf of a galvanic cell?

A. Increasing the concentration of the oxidant at the anode.
B. Increasing the concentration of the oxidant at the cathode.
C. Decreasing the internal resistance of the cell.
D. Raising the temperature of the cell.

More Cells

976. In an electrolytic cell, the cathode is:

A. negative and reduction takes place.
B. negative and oxidation takes place.
C. positive and reduction takes place.
D. positive and oxidation takes place.

977. Which of the following is a possible application of an *electrochemical* cell?

I. A voltage source
II. A pH meter
III. A device for dissolving a gold bar and electroplating the ions onto a nickel bar

A. II only
B. III only
C. I and II only
D. I and III only

978. Which of the following is/are a possible application of an *electrolytic* cell?

I. A voltage source
II. A pH meter
III. A device for dissolving a gold bar and electroplating the ions onto a nickel bar

A. II only
B. III only
C. I and II only
D. I and III only

979. A particular voltaic cell is made with aqueous solutions in both half cells. The emf of the cell depends upon all of the following EXCEPT:

A. the temperature.
B. the concentrations of half cell reactants and products.
C. the specific reactions occurring in the half cells.
D. the volume of solutions used in each half cell.

980. Electrolytic cells are involved in all of the following situations EXCEPT:

A. causing metal to plate onto an electrode.
B. recharging an electrochemical cell.
C. powering an electrical device.
D. decomposing a salt into its constituent elements.

981. A pH meter can be thought of as a hydronium concentration cell. The reference half-cell is kept at a pH of 2.0. The other half-cell consists of a platinum probe placed into the sample being tested. If electrons flow from the reference half-cell into the sample, which of the following is true? (The applicable half-reaction is: $2H^+ + 2e^- \rightarrow H_2$)

A. The probe is the anode and the sample solution has a pH less than 2.0.
B. The probe is the anode and the sample solution has a pH greater than 2.0.
C. The probe is the cathode and the sample solution has a pH less than 2.0.
D. The probe is the cathode and the sample solution has a pH greater than 2.0.

982. Attempts to obtain solid sodium by the electrolysis of aqueous sodium chloride inevitably fail, producing hydrogen gas at the cathode rather than sodium. Which of the following is a possible explanation?

A. Chlorine ions are oxidized more easily than sodium
B. Chlorine is reduced more easily than sodium ions
C. Water is oxidized more easily than sodium
D. Water is reduced more easily than sodium ions

Questions 983-991 use the following table of half-reaction potentials:

Half-reaction	E^0, V
$Ag^+ + e^- \rightarrow Ag$	+0.80
$Cd^{2+} + 2e^- \rightarrow Cd$	−0.40
$2F^- \rightarrow F_2 + 2e^-$	−2.87
$2I^- \rightarrow I_2 + 2e^-$	−0.54
$Na^+ + e^- \rightarrow Na$	−2.71
$Se^{2-} \rightarrow Se + 2e^-$	+0.67
$2H_2O \rightarrow O_2 + 4H^+ + 4e^-$	−1.23
$2H_2O + 2e^- \rightarrow H_2 + 2OH^-$	−0.83
$Zn^{2+} + 2e^- \rightarrow Zn$	−0.76

An electrolytic cell is constructed with a silver and a zinc electrode immersed in the same aqueous solution. The solution is 1*M* in zinc and silver ions. The electrolytic cell is arranged so that the silver electrode is the anode. (Unlike the actual MCAT, questions in this sequence may require you to have answered earlier questions correctly. If you are having trouble, you might want to read the explanations to some of the earlier questions before completing this set.)

983. What is oxidized in this cell?

A. The silver electrode
B. The zinc electrode
C. The water
D. The zinc ions

984. What is reduced in this cell?

A. The zinc electrode
B. The water
C. The silver ions
D. The zinc ions

985. What is the net reaction for this cell?

A. $2Ag^+ + Zn \rightarrow 2Ag + Zn^{2+}$
B. $2Ag + Zn^{2+} \rightarrow 2Ag^+ + Zn$
C. $4Ag^+ + 2H_2O \rightarrow 4Ag + O_2 + 4H^+$
D. No net chemical reaction takes place

986. What is the minimum theoretical voltage necessary to run this electrolytic cell (neglect the overvoltage)?

A. –1.56 V
B. 0 V
C. 0.43 V
D. 1.56 V

987. If the silver electrode were replaced by a sodium electrode, and the silver ions in solution replaced by sodium ions, what would happen?

A. The sodium would be oxidized at the anode and reduced at the cathode, resulting in a layer of sodium metal on the cathode.
B. The sodium would be oxidized at the anode, but zinc ions would be reduced at the cathode. The result would be the replacement of zinc ions by sodium ions in solution.
C. Water would be oxidized at the anode, but sodium would be reduced at the cathode. The result would be the replacement of sodium ions in solution by hydrogen ions, accompanied by the release of oxygen gas.
D. Water would be oxidized at the anode and reduced at the cathode. The result would be the production of oxygen and hydrogen gas.

988. Suppose, instead of using the cell described above, a new cell was constructed out of two half-cells, one made of a zinc electrode in 1 *M* aqueous zinc nitrate, and the other a silver electrode in 1 *M* aqueous silver nitrate. The two electrodes are connected by a salt bridge. An external power supply is attached in such a way as to make the silver electrode the anode. (Assume the nitrate is neither oxidized nor reduced.) Which of the following would most likely occur?

A. Although the salt bridge would slow the process, the net reaction would be the same as before, since the same reagents are involved.
B. Although silver would still be oxidized at the anode, zinc ions would now be reduced at the cathode.
C. Although silver would still be oxidized at the anode, water would be reduced at the cathode, accompanied by the production of hydrogen gas.
D. No reaction would occur, since the resulting cell voltage would be negative.

989. The cell described in the previous question is:

A. a galvanic cell.
B. an electrolytic cell.
C. a concentration cell.
D. a fuel cell.

990. Suppose both electrodes are removed from the original cell and replaced by platinum wires, which can neither be oxidized nor reduced. Likewise, the solution is replaced by pure water. If sufficient voltage is applied, what will the net reaction be?

A. $H^+ + OH^- \rightarrow H_2O$
B. $H_2O \rightarrow H^+ + OH^-$
C. $2H_2O \rightarrow 2H_2 + O_2$
D. No reaction will take place

991. What minimum external voltage is necessary to drive the reaction from the previous problem? (Neglect the overvoltage.)

A. 0 V
B. 1.23 V
C. 2.06 V
D. 2.89 V

Use the Nernst equation in the form given below to help answer questions 992-996.

$$\mathscr{E} = \mathscr{E}^{\circ} - \frac{0.06}{n}\log(Q)$$

992. A concentration cell is created using silver electrodes and aqueous silver solution. The reduction half reaction for silver is:

$$Ag^{+}(aq) + e^{-} \rightarrow Ag(s) \quad E^{\circ} = 0.80$$

What value should be plugged in for E° to derive the E for the concentration cell?

A. 0
B. +0.8
C. −0.76
D. +1.6

993. A concentration cell is created using silver electrodes and aqueous silver solution. The reduction half reaction for silver is:

$$Ag^{+}(aq) + e^{-} \rightarrow Ag(s) \quad E^{\circ} = 0.80$$

If the silver ion concentration at the cathode is 1 *M*, which of the following could be the silver ion concentration at the anode in a galvanic cell?

A. 0.1 *M*
B. 1 *M*
C. 10 *M*
D. A positive emf cannot be achieved because the oxidation and reduction potentials will cancel.

994. A concentration cell is created using zinc electrodes and aqueous zinc solution. The reduction half reaction for zinc is:

$$Zn^{2+}(aq) + 2e^{-} \rightarrow Zn(s) \quad E^{\circ} = -0.76$$

If the zinc ion concentration at the cathode is 1 *M*, which of the following could be the zinc ion concentration at the anode in a galvanic cell?

A. 0.1 *M*
B. 1 *M*
C. 10 *M*
D. A positive emf cannot be achieved because the oxidation and reduction potentials will cancel.

995. Which of the following describes *Q* in the Nernst equation for a concentration cell made with silver electrodes and aqueous silver solution?

$$Ag^{+} + e^{-} \rightarrow Ag \quad E^{\circ} = 0.80$$

A. $[Ag^{+}]_{cathode}/[Ag^{+}]_{anode}$
B. $[Ag^{+}]_{anode}/[Ag^{+}]_{cathode}$
C. $[Ag]_{anode}[Ag^{+}]_{cathode}/[Ag]_{cathode}[Ag^{+}]_{anode}$
D. $[Ag]_{cathode}[Ag^{+}]_{anode}/[Ag]_{anode}[Ag^{+}]_{cathode}$

996. A galvanic cell is created using silver and zinc electrodes, a 1.0 *M* aqueous silver solution, and a 0.1 *M* aqueous zinc solution. The reduction half reactions for silver and zinc are:

$$Ag^{+} + e^{-} \rightarrow Ag \quad E^{\circ} = 0.80$$

$$Zn^{2+}(aq) + 2e^{-} \rightarrow Zn(s) \quad E^{\circ} = -0.76$$

What is the emf for this cell?

A. 1.50 V
B. 1.53 V
C. 1.56 V
D. 1.59 V

---

997. A cup is plated with gold in an electrolytic process. The reduction half reaction for gold is:

$$Au^{3+}(aq) + 3e^{-} \rightarrow Au(s) \quad E^{\circ} = 1.50$$

If the current is 2 amps and the plating lasts for 30 seconds, what mass of gold is deposited onto the cup? (Faraday's constant is 96,500 c/mole e^{-}.)

A. $\dfrac{(197)(2)(30)}{(96{,}500)(3)}$

B. $\dfrac{(197)(2)(30)(3)}{(96{,}500)}$

C. $\dfrac{(197)(30)(3)}{(96{,}500)(2)}$

D. $\dfrac{(197)(3)(96{,}500)}{(2)(30)}$

998. One spoon is plated with gold and another with copper. If a current generator is used to plate both spoons at a constant current, which spoon will acquire 1 gram of plating the fastest? The reduction half reactions are:

$$Au^{3+}(aq) + 3e^{-} \rightarrow Au(s) \quad E^{\circ} = 1.50$$

$$Cu^{2+}(aq) + 2e^{-} \rightarrow Cu(s) \quad E^{\circ} = 0.34$$

A. The copper spoon because copper has a lower emf.
B. The copper spoon because copper reduces with fewer electrons.
C. The gold spoon because gold is heavier.
D. The gold spoon because gold reduces with more electrons.

999. Corrosion is the oxidation of a metal. Iron may be coated with tin, zinc, or another metal to protect it against corrosion. If the surface coat on the iron breaks and the metals are exposed to water, corrosion occurs. The standard reduction potentials for tin, zinc, and iron ions are given below.

$$Sn^{2+}(aq) + 2e^- \rightarrow Sn(s) \quad E^\circ = -0.14 \text{ V}$$

$$Zn^{2+}(aq) + 2e^- \rightarrow Zn(s) \quad E^\circ = -0.76 \text{ V}$$

$$Fe^{2+}(aq) + 2e^- \rightarrow Fe(s) \quad E^\circ = -0.44 \text{ V}$$

Which metal, tin or Zinc, will prevent the corrosion of iron even after the surface coat is broken?

A. Tin, because tin will *oxidize* more easily than iron.
B. Tin, because tin will *reduce* more easily than iron.
C. Zinc, because zinc will *oxidize* more easily than iron.
D. Zinc, because zinc will *reduce* more easily than iron.

Use the following information to answer questions 1000-1001. Iron is corroded via an electrochemical process as shown in the diagram below.

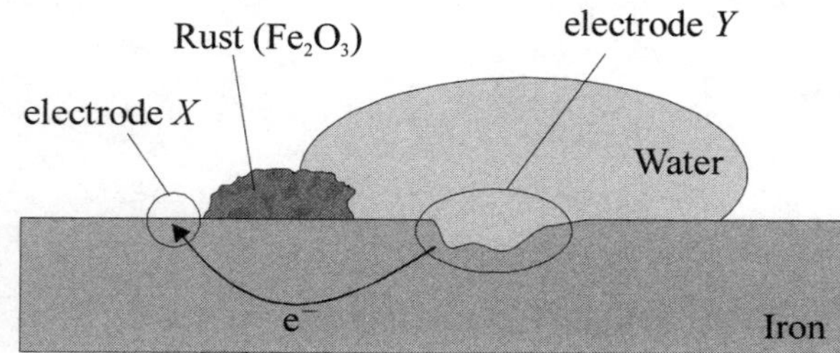

1000. Which of the following describes oxidation in the corrosion process?

A. Iron loses electrons at electrode *Y*.
B. Oxygen accepts electrons at electrode *Y*.
C. Iron loses electrons at electrode *X*.
D. Oxygen accepts electrons at electrode *X*.

1001. Which of the following describes reduction in the corrosion process?

A. Iron loses electrons at electrode *Y*.
B. Oxygen accepts electrons at electrode *Y*.
C. Iron loses electrons at electrode *X*.
D. Oxygen accepts electrons at electrode *X*.

Answers to 1001 Questions in MCAT Chemistry

1. **C is correct.** The rules for significant figures are as follows: 1) Nonzero digits are always significant. 2) Zeros between digits are significant. 3) Zeros preceding all other digits are never significant. 4) Zeros that are both to the left of a decimal point and to the left of all other digits are NOT significant. 5) Zeros at the end of a number with no decimal point may or may not be significant. Numbers written in scientific notation avoid this ambiguity.
2. **A is correct.** The final answer has the same number of significant digits as the measurement with the least number of significant digits. See question 1.
3. **B is correct.** The atomic number is written to the lower left of the elemental symbol. The atomic number is the number of protons.
4. **C is correct.** The mass number is written to the upper left of the elemental symbol. The mass number is the number of protons and neutrons.
5. **C is correct.** The charge is written to the upper right of the elemental symbol. The charge is the number of protons minus electrons.
6. **A is correct.** The mass number is a good approximation of the atomic weight of the element in either amus or grams per mole.
7. **D is correct.** Since Z is the number of protons and A is the number of protons plus neutrons, Z cannot be larger than A.
8. **C is correct.** Charge comes only from protons and electrons. Protons have equal and opposite charge to electrons.
9. **B is correct.** The atomic number is written to the lower left of the elemental symbol.
10. **A is correct.** The mass number is written to the upper left of the elemental symbol.
11. **D is correct.** Since Z is the number of protons and A is the number of protons plus neutrons, A minus Z is the number of neutrons.
12. **C is correct.** The atomic number defines the element.
13. **C is correct.** Atoms with the same atomic number but different mass numbers are isotopes of the same element.
14. **D is correct.** The mass number is a good approximation of the atomic weight of the element in either amus or grams per mole.
15. **C is correct.** The mass number is a good approximation of the atomic weight of the element in either amus or grams per mole.
16. **D is correct.** The mass number is a good approximation of the atomic weight of the element in either amus or grams per mole, so 12 grams is one mole and 24 grams is two moles.
17. **B is correct.** The mass number is a good approximation of the atomic weight of the element in either amus or grams per mole.
18. **A is correct**. The atomic number defines the element.
19. **A is correct.** An amu is precisely defined as 1/12 the mass of an atoms of carbon-12, which is approximately (not exactly) the mass of one proton.
20. **A is correct.** The mass of an electron is less than one-tenth of one percent the mass of a neutron or proton. (Neutrons and protons have approximately, but not exactly, the same mass.) Therefore choice *A*, with the greatest total of protons and neutrons, has the greatest mass.
21. **C is correct**. Elements are *defined* by the number of protons they have (their atomic number). In addition, the question refers to neutral atoms, so the number of electrons is equal to the number of protons.
22. **A is correct.** Looking at the periodic table, the atomic mass of a natural sample of silicon is 28.1. Since that is much lower than the mass of silicon-29 and silicon-30, silicon-28 must predominate.
23. **C is correct.** The periodic table gives the mass of arsenic as 74.9 and the atomic number (number of protons) as 33. Since the mass of both protons and neutrons is approximately 1 amu, the number of neutrons is approximately $75 - 33 = 42$. Note that this is only true because we are told that there is only one naturally occurring isotope…otherwise there could, for example, be a roughly equal mixture of isotopes with 41 and 43 neutrons.
24. **D is correct.** The atomic mass of sodium, according to the periodic table, is 23.0, meaning that the mass of one mole of sodium is 23.0 grams. Thus 3 moles has a mass of 69.0 grams.
25. **C is correct.** Even if you're not sure how to calculate this, you should be able to eliminate choices *A* or *B*; 48 grams is a macroscopic amount of magnesium (an amount you could measure in an undergraduate chemistry lab, for example), and macroscopic amounts of materials contain very large numbers of atoms. From the periodic table, magnesium has a molar mass of 24.3 g/mol, which means that 48 grams of magnesium would contain approximately 2 moles, or $2 \times 6.02 \times 10^{23}$ atoms.
26. **B is correct.** This is a little tricky. I'd do it in this order, although there are other strategies that work: first, how many moles of helium do we have? Since helium has a molar mass of 4.0 g/mol, 2 grams is half a mole. But each mole of helium has 2 electrons, since helium has atomic number 2 and numbers of electrons and protons are equal in a neutral atom. So we have

half a mole of helium multiplied by 2 electrons per mole, giving us one mole of electrons. We are then told that there are 96,500 C/mol, so the answer is *B*.

27. **B is correct.** The charge on one electron is *e*, so one mole is Avogadro's number multiplied by *e*.
28. **A is correct.** The atomic mass of hydrogen, according to the periodic table, is 1.0 amu. Thus, since a single proton has a mass of about 1 amu, the most common isotope contains no neutrons. OR, you could have realized that in acid-base chemistry, hydrogen ions are referred to as "protons," implying there are no neutrons in the nucleus.
29. **A is correct.** The most common isotope of hydrogen has one proton and no neutrons.
30. **D is correct.** Since protons and neutrons both have a mass of roughly 1 amu, and the atomic number of an element represents the number of protons in the element, an isotope with more neutrons than protons would have an atomic mass more than twice its atomic number.
31. **C is correct**
32. **B is correct.** Since naturally occurring lithium has a mass of 6.9 g/mol (see periodic table) and both protons and neutrons have a mass of approximately 1 g/mol, lithium-7 must be predominant.
33. **D is correct**. You must use the periodic table for this question; you need to know that the atomic weight of copper is 63.5. Therefore, the other isotope of copper must be heavier than the predominant copper-63, limiting the choices to C or D. Since more than 2/3 of the copper is copper-63, the atomic weight should be closer to 63 than to the other isotope; that rules out copper-64. (If copper-64 were the other isotope, the atomic weight would have to be less than 63.5.) So, that only leaves D. You *could* confirm this by calculation, but on the actual test, don't waste your time!
34. **B is correct.** In other words, what is the molar mass? Using the periodic table, the answer is $(6 \times 12) + (12 \times 1) + (6 \times 16) = 180$ g.
35. **B is correct.** You need to memorize the name of this family for the MCAT.
36. **B is correct.** You need to know this for the MCAT.
37. **B is correct.** You should know that ions typically form by taking on a noble gas electron configuration. This means losing two electrons for alkaline earth metals.
38. **C is correct.** You should know that ions typically form by taking on a noble gas electron configuration. This means gaining one electron for halogens.
39. **C is correct.** You need to recognize that the B families on the periodic table represent the transition metals.
40. **A is correct.** You need to memorize the name of this family for the MCAT.
41. **C is correct.** The transition metals are more likely to form more than one possible ion because they have *d* orbitals that can hold or lose electrons.
42. **A is correct.** All alkali metals (first column of the periodic table) form 1^+ ions, and all halogens form 1^- ions, so the salts they form will be one-to-one.
43. **A is correct.** Malleability is a metallic characteristic. Metallic character increases to the left and down.
44. **D is correct.** Electronegativity increase to the left and up. You should know that fluorine has the greatest electronegativity of all elements.
45. **B is correct.** Electronegativity is higher for non-metals than for metals; it thus increases as you move up and to the right on the periodic table. Noble gases, however, form a special case. They should *not* be considered the most non-metallic elements, but rather as having special properties all their own. The most non-metallic elements are the halogens, thus the answer is *B*.
46. **A is correct.** This is definitional.
47. **A is correct.** Accepting electrons is a characteristic of non-metals. See question 45 for a discussion of how to determine the most non-metallic element in a set.
48. **A is correct.** This question is really asking for the highest electronegativity, which is a characteristic of non-metals. See question 45 for a discussion of how to determine the most non-metallic element in a set.
49. **C is correct.** Size increases down and to the left on the periodic table; it can be thought of as a property that is larger for metals.
50. **A is correct.** Negative ions are *much larger* than their neutral counterparts; positive ions are *much smaller*. For a more rigorous argument, note that all four choices are isoelectronic (have the same number of electrons). Thus the choice with the smallest number of protons should be the largest.
51. **B is correct.** Positive ions are *much smaller* than their neutral counterparts; negative ions are *much larger*. Since the size of neutral atoms decreases as you move from left to right across the periodic table, neutral aluminum is bigger than neutral sulfur, and the correct answer is *B*.
52. **B is correct.** The attraction of the nucleus on the outer most electron is related to the ionization energy which increases to the right and up on the periodic table.
53. **D is correct.** Electronegativity is a measure of how strongly an element attracts electrons in a bond, while the dipole moment of a bond depends on whether the electrons are pulled to one side or other of the bond. In order to have a large dipole moment, one element must be pulling much harder than the other. (Note: dipole moment also depends on the length of the bond.)

54. **D is correct.** As electrons are removed from an atom, each successive electron becomes harder to remove.
55. **D is correct.** Since atoms get smaller as you go the right across the periodic table, but get larger as you go down, the size is not likely to change much as you go down and to the right (or up and to the left). This "diagonal relationship" is not an unbreakable rule (selenium is a bit bigger than phosphorous, as it turns out), but it is a good basis for estimates.
56. **C is correct.** First of all, note the position of the elements on the periodic table: they are all in the same period. (Also, they are all solids—no easy out this time!) So this is probably an issue of periodic trends, and you can narrow it down to *C* or *D* (the two extremes). Density is mass/volume; mass of course increases to the right, and volume decreases, so density increases to the right. (There is some small effect due to the crystal structure adopted, but it's not likely to be enough to compensate for the size and mass difference between aluminum and sulfur atoms.)
57. **C is correct.** Fluorine's a gas, and the others are solids—that trumps the effects discussed in 56.
58. **C is correct.** Throw out the carbon, because it's a solid. The volume of an atom doesn't matter for a gas (gases are mostly the empty space between atoms), but the mass still does.
59. **A is correct.** Based on the logic of the previous question, you might think the answer should be *B*. But be careful: chlorine is a diatomic gas and has a molar mass of 71 g/mol, while argon is monatomic and has a molar mass of 40 g/mol.
60. **C is correct.** The second ionization energy of lithium is much, much larger than the first ionization energy because after one electron is removed the resulting ion has a noble gas configuration. Element X does not have nearly the same jump, so it is not an alkaline metal. Since element X does have a similar first ionization energy to lithium, however, it is a metal.
61. **B is correct.** The best bet for chemical similarity is to stay in the same column of the periodic table.
62. **B is correct.** A family, also known as a group, is a column of the periodic table. (Rows are called periods.)
63. **D is correct**. The best bet for chemical similarity is to stay in the same columns of the periodic table. K is in the same column as Na, and Br is in the same column as Cl.
64. **B is correct.** Na^+ has a noble gas configuration; it is thus very hard to remove an additional electron.
65. **A is correct.** Cl is the most non-metallic of the elements, and thus has the highest first ionization energy. It is considerably easier to remove an electron from Cl^{2-}, since it has more electrons for the same amount of positive charge.
66. **C is correct.** This question is equivalent to asking which element has the greatest first ionization energy. First ionization energy is higher for nonmetals than for metals, and thus increases as you move up and to the right on the periodic table.
67. **D is correct.** Electronegativity measures the attraction of an element for electrons within a chemical bond.
68. **D is correct.** Solutions containing ions of transition metals are frequently colored.
69. **D is correct.** If atom A has a greater radius than atom B, then it is more metallic. Metals have lower electronegativities and first ionization energies than nonmetals.
70. **D is correct**. This one is tricky. Catching fire when exposed to water suggests extreme reactivity, probably due to the ease with which alkali metals lose their electrons. So we want a metal that loses its electron even more easily, which means travelling down the periodic table (or, in general, to the left, but that's not an option in the first column!). Thus, potassium.
71. **C is correct.** Choice A is factually wrong; positive ions are smaller than the neutral atom. Choice B is at best irrelevant; removing additional electrons from a positive ion is always harder than removing the first electron. This leaves C and D. Choice C suggests that each outer electron is more tightly held in Fe^{2+} than in Fe, which is true and which explains the increased ionization energy in Fe^{2+}. Choice D indicates that Fe^{2+} is higher in energy than Fe, which is true but does not address the issue of how hard it is to remove an additional electron from Fe^{2+}.
72. **C is correct.** Mg^{2+} is isoelectronic with (has the same number of electrons as) Na^+. Thus choice *A* is factually incorrect. Mass is almost irrelevant to chemical behavior (that's why different isotopes of the same element still behave similarly). *C* is correct: Mg^{2+} has more protons for the same number of electrons.
73. **B is correct.** Two electrons are required to form a typical covalent bond.
74. **A is correct.** Bond energy is closely related to bond dissociation energy, the energy necessary to break the bond. Bond length is inversely related to bond strength. So long bonds have low strength and low bond energy.
75. **B is correct.** Since bond formation decreases entropy, it is always exothermic. Thus breaking a bond is always endothermic, and requires energy.
76. **C is correct.** Both positively charged nuclei pull on the electrons in the bond, like a tug of war.
77. **C is correct.** The force holding the atoms together is electrostatic.
78. **A is correct.** The positively charged nuclei repel each other, while they each attract the bonding electrons. An equilibrium is established at the bond length.
79. **B is correct.** This is a point that often causes confusion. A double bond comprises a sigma bond and a pi bond, while a single bond has only a sigma bond. Pi bonds are weaker than sigma bonds, but of course are better than nothing. Thus a double bond (sigma and pi) is stronger than a single bond (sigma only) but less strong than two single bonds (sigma and sigma). Put another way, double bonds have less bond energy *per electron* than a single bond.
80. **A is correct.** All four choices represent true statements, but only choice A has anything to do with the amount of electron density around oxygen.
81. **C is correct.** $14 \div (14 + [2 \times 16])$.

82. **D is correct.** CO has a triple bond, while CO_2 and K_2CO_3 both contain carbon-oxygen double bonds. The CO bond in CH_3OH is a single bond, which is the longest of the three.
83. **D is correct.** Resonance spreads the double-bond character equally among all three C—S bonds.
84. **D is correct.** It is good to have answers "compete" against each other in questions like this. For example, *B* easily beats *A*, since they have the same amount of carbon but choice *A* has much heavier non-carbon atoms. But choice *C* beats *B*, because choice C has almost the same mass of non-carbon atoms as choice *B*, but has much more carbon. Finally, choice *D* just nicks out choice *C*; they have identical masses of carbon, but choice *D* has a slightly lower mass of non-carbon (9 x 1 + 14 vs. 8 x 1 + 16).
85. **B is correct.** *A* is, in fact, a common industrial chemical. The fluorines in SF_6 do not have expanded octets; each is bonded only to the sulfur. The sulfur has an expanded octet, but it has empty d-orbitals available. As far as choice *C* goes, hydrogen can form a negative ion, such as in sodium hydride, NaH. And lead, despite the fact that it is not a transition metal, does have two common valences: +2 and +4.
86. **C is correct.** In this case you do have to draw a Lewis structure.
87. **C is correct.** Consider, for example, NaCl. It violates choices *A*, *B*, and *D*, as do all ionic compounds.
88. **B is correct.** This is essentially a question about units. Grams and moles are macroscopic units, meaning they are appropriate for an amount of material you could measure out in a freshman chemistry laboratory (or a kitchen!). Only *amu* is an atomic-scale unit. If the mass of a single molecule of water were measured in grams, the number would be very small (something like 10^{-22}).
89. **B is correct**. Here's the conversion with units:

    18 g/mol x 1 mol/6.02 x 10^{23} molecules x 1 kg/1000 g

    Watch the MCAT on the grams-kilograms conversion...it sneaks in there sometimes.
90. **D is correct.** Carbon tetrachloride is CCl_4. The atomic mass is about 12 + 4 x (35.5) = 154. Of that, the chlorine is 4 x (35.5) = 142. So the percent by mass is 142/154 x 100. This is certainly more than 80%, so the answer is D. There is no need on the test to calculate problems like this exactly; it's enough to say answers A, B, and C are too small, so D must be correct.
91. **B is correct.** Often, the easiest way to do problems like this is to work backwards: go through each answer choice and see which matches the given information. To save more time, calculate one element at a time. In other words, consider choices *A* and *B*; the only difference is the amount of hydrogen. What is the percent by mass of hydrogen is $ClCH_3$? The total mass is 35.5 + 12 + 3 x 1 = 50 or so. The hydrogen is 3 x 1 = 3. 3/50 is about 0.06, or 6%. That's too much hydrogen; eliminate choice A. For choice B, the total mass is still about 50 (one hydrogen doesn't make much difference), but the mass of the hydrogen is 2 x 1 = 2. And 2/50 is about 0.04, so the 4% hydrogen checks out. Choice C has a total mass of 35.5 + 2 x 12 + 5 x 1, or about 65. Since there are five hydrogens, the percent by mass is about 5/65, which is bigger than 4%, so C is out. Finally, choice D has a total mass of 2 x 35.5 + 12 + 2 x 1 = 85. Two hydrogens gives a percent by mass of 2/85 for the hydrogen, which is less than 4%. So choice *B* must be correct. Of course, there are many other ways to approach this problem. Use whatever method works for you!
92. **C is correct.** A perfect opportunity to "cheat." (Don't worry--AAMC doesn't hold it against you!) We know the molecular weight is approximately 99, and we can determine the molecular weights of the choices easily…unlike percent by mass, there is no division involved. Choices *A* and *B* are way too light. Choice *C* looks about right: 2 x 35.5 + 2 x 12 + 4 x 1. (Don't you dare do this to four significant figures! We're using process of elimination, so you don't have to be sure the right answer is *exactly* right.) Choice *D*, with one more chlorine than choice *C*, is way too high.
93. **D is correct.** A particularly easy one for working backward. Choice *A* has an oxygen fraction of 16 ÷ (16+31), or about 30%…too low. Choice *B* has an oxygen fraction of 32 ÷ (32+31), which is very close to 50%…just a bit too low. Choice *C* is 48 ÷ (48+62), which is again too low. Finally, choice *D* is 80 ÷ (80+62), which is well above 50% and must therefore be the correct answer.
94. **D is correct.** "Cheating" time again! Just use the molecular weight. Choice *C*, for example, has a molecular weight of 3 x 31 + 7 x 16 = 205…too low. Choice *D* is the only one with a higher molecular weight and therefore must be correct.
95. **C is correct.** Water was 1.5 grams of the hydrate (7.0 – 5.5). That gives a fraction of water in the hydrate of 1.5/7.0 or a bit more than 20%.
96. **B is correct.** Empirical formulae are always in lowest terms; the empirical formula for choice *A*, for example, would be NO_2.
97. **B is correct.** One mole of carbon dioxide has a mass of 12 + (2 x 16) = 44 grams. So 22 grams must be half a mole.
98. **D is correct.** One mole of NH_3 has a mass of 14 + (3 x 1) = 17 grams. So 34 grams must be 2 moles.
99. **B is correct.** Three moles of water vapor were produced, since 54 grams of water are 3 moles. Thus, one mole of C_2H_5OH was consumed. Since C_2H_5OH contains 6 hydrogens, 6 moles of hydrogen participated in the reaction.
100. **B is correct.** Choice *A* is much higher than 25% nitrogen by mass, since an atom of nitrogen has a much greater mass than one atom of hydrogen. Choice *B* is very close; let's check the others. Choice *C* has way too much nitrogen; there's basically one nitrogen for every CH_3, or something like 50% nitrogen by mass. Choice *D* also has too much nitrogen.

101. A is correct. In other words, one out of four atoms are nitrogen.

102. D is correct. Two nitrogen atoms have a mass of 28. If that is 10% of the compound, the mass of the compound is 280.

103. A is correct. Empirical formulae are always "reduced to lowest terms." In other words, the ratio of carbon to hydrogen in this compound is 6:6, which is 1:1.

104. B is correct. First, gather together atoms of the same element: there are two carbons, four hydrogens, and two oxygens. Now reduce; we can divide by two to arrive at choice B.

105. A is correct. Work this one in reverse; check the answers one at a time. Choice *A* has a hydrogen fraction of $1 \div (1 + 16)$, which seems as if it could be right. Choice *B* has twice the hydrogen, which would give it over 10% H. Choices *C* and *D* are not empirical formulae!

106. D is correct. The molecular weight alone is not sufficient; two different compounds might coincidentally have very similar molecular weights. Likewise, the percent by mass alone is not sufficient; it leads only to the empirical formula. If both are known, however, the empirical formula can be multiplied by a suitable constant to arrive at the correct molecular weight. Using that constant to multiply the coefficients in the empirical formula will give you the molecular formula.

107. C is correct. The molar mass cannot be determined from the empirical formula, since the molecular formula is unknown (it might, for example, be C_5H_{10}). The percent composition can be found, however, since this only depends on the relative amounts of the elements present. This is exactly the information the empirical formula provides.

108. D is correct. The volume of the block is $3 \times 5 \times 8 \text{ cm}^3 = 120 \text{ cm}^3$. But we need an answer in m^3. So $120 \text{ cm}^3 \times (10^{-2} \text{ m}/1 \text{ cm})^3 = 1.2 \times 10^{-4} \text{ m}^3$. Finally, the density is $2.0 \text{ kg}/1.2 \times 10^{-4} \text{ m}^3$. It is not necessary to divide the 2.0 by 1.2; the exponent of 10^4 guarantees that D is the correct answer.

109. D is correct. Speed is distance over time, so the distance must definitely be in the numerator; this rules out choice A. Because we want meters rather than kilometers, we must divide by 0.001 which is the same as multiplying by 1000; therefore choice D is correct. The (365)(24)(60)(60) is the number of seconds in a year; a year being the amount of time it takes the Earth to orbit the sun. This kind of grade school scientific fact sometimes shows up on the MCAT, but if your general science background is very weak, don't panic; there's frequently another way to figure out the answer, as in this case. Oh, and the (2)(3.14)? The circumference of a circle is $2\pi r$, another miscellaneous grade school fact.

110. A is correct. Phase changes are considered physical changes.

111. A is correct. Intermolecular bonds can be broken during physical changes, but the molecules themselves should not change.

112. D is correct. Intramolecular bonds are broken and formed, therefore it is not a physical reaction.

113. D is correct. Choices B and C aren't even balanced. (Technically, choice A isn't balanced either, because the charge on the left isn't the same as the charge on the right, but we'll get to that in lecture 7.) Methane is CH_4, so that pretty much limits us to choice D. Also, combustion means combination with oxygen, so that also eliminates choices A and C.

114. C is correct. You can either balance the equation yourself, or just check which equation is balanced. The second technique is probably quicker: the carbon is not balanced in choice A, the hydrogen is not balanced in choice B, and the hydrogen is not balanced in choice D. There are other elements out of balance in the incorrect choices, but it suffices to find one!

115. D is correct. All of the reactions are properly balanced; the problem is in the nomenclature and ions. The "(II)" in copper (II) chloride means copper has a charge of +2. Chloride, being a halogen, has a charge of –1. So copper (II) chloride is $CuCl_2$. Likewise, carbonate has a charge of –2, so iron (II) carbonate is $FeCO_3$.

116. C is correct. First, balance the reaction: $C_{12}H_{22}O_{11}(l) + 12O_2(g) \rightarrow 12CO_2(g) + 11H_2O(g)$. From there, the stoichiometry is easy.

117. C is correct. Again, balance the reaction: $C_6H_{12}O_6(s) + 6O_2(g) \rightarrow 6CO_2(g) + 6H_2O(g)$.

118. D is correct. The balanced reaction: $2C_6H_{14}(g) + 19O_2(g) \rightarrow 12CO_2(g) + 14H_2O(g)$.

119. B is correct. First, balance the reaction: $2Fe(s) + 1.5O_2(g) \rightarrow Fe_2O_3(s)$. Note that although balanced equations are "supposed" to have integer coefficients, it is not necessary for doing stoichiometry. According to the reaction, 2 moles of iron should reaction with 1.5 moles of oxygen…if there are two moles of oxygen, there will be 0.5 moles left over.

120. A is correct. First of all, the limiting reagent must be a reactant, so the answer must be *A* or *B*! Next, balance the reaction: $Au_2S_3(s) + 3H_2(g) \rightarrow 2Au(s) + 3H_2S(g)$. According to the balanced reaction, one mole of $Au_2S_3(s)$ reacts with three moles of hydrogen gas. If there are five moles of hydrogen gas, then there is more than enough hydrogen, and the $Au_2S_3(s)$ will run out first.

121. D is correct. A scary question, because of the extra information. I'd be worried about a trick (is this a limiting reagent problem is disguise?). So, to check whether this is a trick question of some sort, see if the equation does predict 15 moles of $N_2O_4(l)$ would give 36 moles of water. Multiplying the equation through by 3 does give this ratio, so the reaction *is* stoichiometric…no trick here, just superfluous information. Multiplying the equation through by 3 gives 12 moles of $N_2H_3(CH_3)(l)$.

122. B is correct. Now we explicitly have a limiting reagent problem. 10 moles of $N_2O_4(l)$ should have produced 24 moles of water, but did not. Therefore, the other reactant must have run out first.

123. C is correct. The coefficients in balanced reactions refer to moles or molecules, not grams. Choice *D* is correct, because it represents 1 mole of nitrogen gas, 3 moles of hydrogen gas, and 2 moles of ammonia gas.

124. **B is correct.** According to the balanced reaction, 11 moles of hydrogen gas should react with 11/3 = 3.7 moles of nitrogen gas. Since we started with 4.5 moles of nitrogen gas, we should finish with 0.8 moles, which is not a choice! So this is a trick question: the reaction must not go to completion. We are also given the moles of ammonia: six moles of ammonia require 3 moles of nitrogen, and we started with 4.5; thus, 1.5 remain. This question illustrates a good general principle about MCAT questions: trick questions will sometimes reveal themselves by not having the answer you expect as a choice.

125. **C is correct.** Hydrogen is the limiting reagent. Eleven moles of hydrogen should produce 11 x 2/3 = 7.3 moles of ammonia. If it actually produces 6 moles, the yield is 6/7.3, which is a lot more than half.

126. **D is correct.** Just a long, annoying stoichiometry problem. Remember that although you have an *average* of a bit over a minute a question of the MCAT, you will answer some questions in 20 seconds, which means you can take more time on others. Three minutes is not unreasonable for this kind of question. Now the details: HCl has a molar mass of 36.5 g/mol, so 328 grams is 9 moles. (You can estimate this calculation; a good approach is to look for a round number. For example: 36 goes into 100 around 3 times, so it goes into 300 about 9 times. Nine moles should be a bit over 300 because 3 times 36 is more than 100. Nine looks good, so use it.) Nine moles is 75% of what the full yield should have been, so we had enough PCl_3 to produce 12 moles of HCl. By applying the mole ratios, we find that 12 moles of HCl would be produced by 4 moles of PCl_3.

127. **A is correct.** Phosphorous trichloride has a molar mass of 31.0 + 3 x 35.5 = 137.5 g/mol. 275 grams is thus very close to 2 moles. From the stoichiometry in the balanced reaction, we find that ½ mole of $P_4(s)$ is needed. Each mole of $P_4(s)$ has a mass of 4 x 31.0 = 124 grams, so the correct answer is *A*.

128. **B is correct.** Since each mole of arsenic gas has a mass of 4 x 74.9 grams, 75 grams must be very close to 1/4 mole. The stoichiometry indicates six times as many moles of carbon are needed, or 6/4 = 1.5 moles of carbon. Since each mole of carbon has a mass of 12 grams, we find that 18 grams of carbon are required.

129. **B is correct**. Straightforward stoichiometry: the ratio is 1:2.

130. **A is correct.** The atoms in solid ice form a regular pattern. One piece of evidence for this fact is that ice has a well-defined melting point.

131. **D is correct.** Ionic compounds, for example, do not contain molecules.

132. **C is correct.** Ionic compounds generally consist of a metal and a nonmetal (or, occasionally, a polyatomic cation like ammonium). HCl is a molecular substance; indeed, it is a gas at room temperature. The fact that it splits into ions when placed in water does *not* indicate definitively that it is an ionic compound in its pure form.

133. **D is correct.** The key word is *molecular*. Although all the other answers represent crystalline solids, they are of the network, ionic, and metallic types respectively.

134. **C is correct**. You may have been looking for hydrogen bonding, which would probably be the best answer, but it's not there. Polarity is a prerequisite for hydrogen bonding, so choice *C* is the best alternative.

135. **C is correct**. You don't need to know about the structure of diamond to answer this question; you can solve it using process of elimination. Nevertheless, it is helpful to know that diamond is one of the best-known examples of a **network covalent solid**, a solid where all the bonds are covalent (no intermolecular bonds; in fact, no distinct molecules). Choice *A* is nonsense: atoms don't provide strength, bonds do. Choice *B* is irrelevant; solid oxygen can be formed under very high pressure, and it certainly doesn't have a high melting point! Choice *D* would lead to diamond having a *lower* melting point, if diamond were a molecular solid. So not only is choice *D* incorrect, it also gives us a clue that diamond is *not* a molecular solid. Thus, process of elimination leaves us with choice *C*.

136. **B is correct**. These compounds are ionic (each containing a metal and a nonmetal). Ionic compounds, like network covalent compounds, don't have individual molecular units; choice *C* is therefore inapplicable. Choice *A* is factually wrong (fluorine is to the right, and therefore more electronegative), and choice *D* is also factually wrong (fluorine is more massive than oxygen). Choice *B* is reasonable: magnesium is 2+ and oxygen 2–; the electrostatic attraction should be larger than that between the similarly sized but less charged sodium (1+) and fluoride (1–) ions.

137. **D is correct.** n, l, m_l, and m_s.

138. **A is correct.** This is just a definition.

139. **A is correct.** Energy depends primarily on *n*.

140. **C is correct.** *l* must be less than *n*. This rules out choice *A*. m_l must be less than or equal to *l*; this rules out choices *B* and *D*. The remaining choice is *C*.

141. **D is correct**. For n=3, we have *s*, *p*, and *d* orbitals. Two electrons can fit in the *s* orbitals, six in the *p* orbitals, and ten in the *d* orbitals. (The easiest way to see this is to look at the blocks on the periodic table.) The total is therefore 18. It is true that the $n = 4$ shell starts filling before the $n = 3$ shell is complete, but this point is irrelevant to the question.

142. **C is correct.** If you thought the answer is *D*, you should realize that this is not an experimental issue…there is no precise orbit to determine!

143. **B is correct.** Realize that the Heisenberg uncertainty principle relates position and momentum, not position and velocity. Thus objects with large mass can have smaller uncertainties in their momenta than smaller mass objects with the same position uncertainty. This is why a nucleus can be localized much more accurately than an electron.

144. C is correct. In fact, the formula is E = $-13.6\ eV/n^2$, although you don't need to memorize that for the MCAT. Choice *A* is not true, because the atom could be in an excited state. Choice *D* is not true, because the charge on the nucleus is greater in the He^+ ion.

145. C is correct. Second period elements cannot have more than four atoms attached, since they have no empty *d* orbitals.

146. B is correct. Use process of elimination: *C* is false; you can even ionize an atom without affecting its nucleus. *A*, although true as far as elements in their ground state are concerned, says nothing about excited states. It is difficult to see what would make choice *D* correct. But choice *B* agrees with something that you should know about orbitals (higher *n* means greater size), and gives a plausible explanation for the observations.

147. D is correct. Remember that to be absorbed, the photon must have exactly the right energy to boost the electron from one level to another, or enough energy to ionize the atom. It's also important to realize the question asks for the photons that could be absorbed by an electron in its ground state, so we are starting at the –14.9 eV (or lowest-energy) level. From there it takes 6.7 eV to raise the electron to the next level, 9.6 eV for the level after that, and so on. This suggests choice *D*. The last two entries in choice *D* (15.0 eV and 16.1 eV) have enough energy to ionize the atom, and are thus allowed.

148. A is correct. It could fall to the intermediate level, giving off an energy of $4.1-2.3 = 1.8$ eV.

149. C is correct. The ground state configuration of sulfur is [Ne] $3s^23p^4$. According to Hund's rule, the *p* orbitals fill up separately first, then start to pair. So the first three *p* electrons go into different orbitals, and the fourth doubles up, leaving two still unpaired.

150. C is correct. The first row of transition metals is 3*d*.

151. D is correct. It doesn't matter in what order the terms are written.

152. C is correct. The iodide ion has one more electron than iodine, giving it a noble gas configuration.

153. D is correct. There's a special rule for positive ions: remove the electrons with the highest principal quantum number first. Since the ground state of the chromium atom is [Ar]$4s^23d^4$, then according to the rule, the first three electrons come from the 4*s* orbitals, and only then are the electrons in the 3*d* orbitals removed.

154. A is correct. In the ground state of an atom, the 2*s* subshell should be followed by the 2*p* subshell. If this is the configuration of a neutral atom, it represents an excited state of boron.

155. A is correct. In the ground state, the 3*p* orbitals fill before the 3*d* orbitals.

156. D is correct. Ions often have the same electronic configuration as neutral atoms of a different element (for example, the fluoride ion is isoelectronic with the neon atom). This is not necessarily true, however, and if excited states are allowed, any electron configuration that does not violate the rules of quantum numbers could be an ion.

157. C is correct. Now three electrons could go into each orbital.

158. C is correct. Although the kinetic-molecular theory defines temperature in terms of the *average* speed of the molecules, it allows for (actually guarantees) variations.

159. D is correct. In fact, in the kinetic-molecular theory of gases, the volume of each gas molecule is assumed to be zero.

160. C is correct. All else being equal, a faster molecule has more momentum, and thus exerts more force on the wall. Likewise, more collisions means more force (on average). But the volume of a molecule is irrelevant to the average force; volume does not enter into Newton's laws!

161. B is correct. Pressure and temperature are directly proportional.

162. C is correct. Pressure and volume are inversely proportional.

163. D is correct. In any gas, molecules have a wide range of speeds. In a gas at 300°C, the *average* speed of a molecule will be greater than in a gas at 150°C, but the spread will be very wide in both cases.

164. B is correct. The ideal gas law states that PV=nRT. Using proportional reasoning, if the temperature is increased by a factor of 1.5 and the pressure is increased by a factor of 1.5 (with the number of moles constant) then the volume is unchanged.

165. A is correct. You could use the ideal gas law, but it's a lot quicker to realize the conditions described are fairly close to STP. Under these conditions, one mole of gas should occupy about 22 L, and thus half a liter should comprise about 1/40 of a mole. Of course, if the answer choices had been closer together, we would have had to be more careful.

166. A is correct. Use the ideal gas law, PV=nRT. If the volume is cut in half, then the number of moles will be cut in half. In addition, the pressure is dropped to 90% if what it was, so the number of moles should drop by this factor as well. With the temperature constant, this gives a new number of moles which is 45% of the old amount. Since mass is proportional to number of moles for a given gas, the new mass is 45% of 20 grams, or 9 grams.

167. D is correct. Try process of elimination on this one. As always with a "why" question, you can eliminate choices either because they are factually incorrect or because they aren't an explanation for the "why." Choice *A*, for example, could be an explanation, but how is xenon going to have argon as a byproduct? Radioactive decay generally occurs through emission of a small particle like an alpha, beta, or positron particle, but argon has an atomic number 36 less than xenon! Choice *B*, again, could make sense, but all of the noble gases are extremely stable (except for radon, which is radioactive). Choice *C* is nonsense—it makes it sound like argon is usually found in compounds. Finally, choice *D* is factually true, since argon has a similar molar mass to nitrogen, and makes sense, since lighter gases such as helium travel with enough velocity to escape from the atmosphere.

168. B is correct. When using proportional reasoning, be careful to make sure that all other factors are constant. Although *D* is a true statement, it's technically only true when temperature and number of moles are both constant.

169. B is correct. A sneaky loophole, and probably not one you would have thought of before seeing the answer choices. Always read all answer choices before making your selection!

170. C is correct. A simple consequence of PV=nRT. Once you plug in the numbers for both situations, you see that volume must be different, so they could not be stored in identical containers.

171. C is correct. The pressure must be tripled to 3 x 850 torr = 2550 torr. That means the additional pressure needed is 2550 – 850 = 1700 torr.

172. D is correct. According to the ideal gas law, the new temperature should be triple the value of the original. But be careful! Temperatures must be expressed in Kelvins you're using the ideal gas law. So 25°C is 298 K, which, when tripled, is almost 900 K, or about 600°C. Alternatively, you might realize that tripling the pressure at constant volume and number of moles should increase the temperature a lot.

173. D is correct. We have to be told what happened to the temperature!

174. D is correct. According to the ideal gas law, both pressure and volume are directly proportional to (absolute) temperature.

175. A is correct. Fifty grams of oxygen is a considerably smaller number of moles than fifty grams of hydrogen, because the molecular weight of oxygen is so much larger. Therefore, since pressure is directly proportional to the number of moles, the pressure of the hydrogen is higher.

176. C is correct. All of the parameters in the ideal gas law are the same, so the number of moles (and thus molecules) is the same. This narrows the choices down to *C* and *D*. The mass, however, is different, since the molar mass of oxygen is higher. Thus, the density is also different. Any answer with choice II in it must also be wrong, eliminating choice *D* and leaving only *C*.

177. D is correct. There is no difference between He and Ne when we're doing an ideal gas problem of this kind, since pressure, volume, temperature, and number of moles are all known quantities. The identity of the gas would only cause a problem if we wanted to convert from moles to mass or vice-versa, but mass is not involved in this problem.

178. B is correct. The total pressure is 760 torr at STP, so the partial pressure of the nitrogen must be 760 – 200 – 10 – 8 = 542 torr.

179. A is correct. Don't waste your time trying to work out all the stoichiometry. On a problem like this, where the calculation is spelled out for you, it is far better to look at what is different among the answer choices. The only differences have to do with the first factor (is it 1000 or 0.001); and the 0.35 (should it be in the numerator or denominator?) The 0.35 must be the percent yield. If percent yield is less than 100%, we know that the actual product is only a fraction of what is predicted by the equation; thus the percent yield must reduce the amount of product. Multiplying by 0.35 makes a number smaller; dividing makes it larger. Knowing this, we see that we need to multiply, narrowing the choices to *A* or *C*. The 1000 is probably the conversion from milliliters to liters. The 24.5 mL we are given can be converted to liters by multiplying by the conversion factor 1 L/1000 mL. Dividing by 1000 is the same as multiplying by 0.001. The correct answer must be *A*.

180. B is correct. At STP, we know that 22.4 L comprises one mole. So 22.4 L/mol x 1.25 g/L gives about 28 g/mol, suggesting that CO is the gas.

181. C is correct. According to the ideal gas law, if the pressure, temperature, and volume are all the same, then the number of moles must be the same too. Looking at the answer choices, let's write a balanced equation for each. Choices *A* and *B* give $2CH_4 + 3O_2 \rightarrow 2CO + 4H_2O$. In this case, the number of moles of gas increases, and the pressure would change at constant temperature and volume. Whether or not one gas is in excess if irrelevant; at least some of the reactants would still be converted into products, raising the number of moles of gas. Choice *C* gives a balanced equation of $CH_4 + 2O_2 \rightarrow CO_2 + 2H_2O$, which has the same number of moles of gas on each side, and is thus probably right. Finally, choice *D* is chemically implausible.

182. B is correct. First, realize that 64.0 grams of oxygen is two moles (O_2 is diatomic). If you want, just plug into PV=nRT. (Round a lot, though; the answer choices are quite far apart and a detailed calculation is a waste of time!) Or you can estimate by comparing to STP: at STP, two moles of gas would occupy 22.4 x 2 ≈ 50 L. So 3.0 L is less than 10% of that. The 5.0 atmospheres pressure helps pack things down, but that's only a factor of 5. So the temperature also must be somewhat lower than STP. Personally, I think this is one problem where plug & chug might be easier, but it's up to you.

183. A is correct. At STP, the total pressure is 1 atmosphere, or 760 torr. The partial pressure is just the mole fraction multiplied by the total pressure, or 0.8 x 760 torr.

184. A is correct. The mole fraction is the ratio of the partial pressure to the total pressure. Since 5 atm = 5 x 760 torr, we calculate a mole fraction of 35/(5 x 760).

185. D is correct. Consider the ideal gas law, PV = *n*RT. We know P, R, and T, but we know neither V nor *n*. If we were given the volume, we could find *n* and thus the mass, but as it is we're out of luck (or in luck—no calculations to do!).

186. D is correct. We can't know the percent by mass without knowing the identity of the other constituents.

187. B is correct. The mole fraction of the carbon dioxide is 30/760, or about 0.04. But since carbon dioxide has a somewhat higher molar mass than nitrogen, it will have a somewhat higher percent by mass than mole fraction. That's enough to choose

B (*C* is way too high). If you actually want to calculate it, suppose there are 100 moles of gas total (the percentage won't depend on the number of moles present, and 100 is easy to work with). Then there's 4 moles of carbon dioxide and 96 moles of nitrogen. The total mass would be (4 x 44) + (96 x 28), and the fraction of carbon dioxide by mass would be 4 x 44 / ([4 x 44] + [96 x 28]). This calculation is kind of a pain, though; it's probably easier to just reason through it.

188. B is correct. There are many ways to approach this problem. First of all, the mole fraction is 30/760 = 0.04. Imagine there were 100 moles of gas present. Then there would be 4 moles of carbon, with a mass of 4 x 44, or 176 grams. That's 10% of the total mass, so the other species must have nine times that mass, or about 1600 grams. Since the other species would have 96 moles present, its molar mass would be 1600/96, or about 17 g/mol. Of the choices, only methane (CH_4) is even close.

189. B is correct. The ideal gas law works with partial pressures as well as total pressure. At 1 atmosphere and 0°C, 22.4 L would be one mole. Compared to the STP values, the volume and pressure are both cut in half, and the temperature is slightly higher, so the number of moles should be *about* 1/2 x 1/2, or 1/4. One-fourth of a mole of nitrogen gas (N_2) is about 7 g.

190. C is correct. This question illustrates the risk of ignoring the restrictions imposed on equations. You might think that Boyle's law leads us to choice *B*, but Boyle's law only applies if temperature and number of moles are constant. We are not sure about the temperature here, and the number of moles certainly is not constant, because some gas was allowed to escape. If we read our choices, however, the solution may become apparent. Choice *C* is correct: the weight on the piston does not change, so the force per area, or pressure, on the top of the piston is constant. Since this pressure must be matched by the gas pressure in order to support the piston, the pressure in the container must be the same for both recordings.

191. B is correct. If you didn't already know this (and you should, for the MCAT), you could use process of elimination by looking at units. Choice *A* gives g/mol x m^6—clearly not energy! Choice *B* gives mol x J/mol K x K, which works. Choices *C* and *D* give the units of force, so they're out. This kind of dimensional analysis can often help narrow down the answer choices on a formula-type question.

192. D is correct. All of the experiments described depend on the *mass* of the gas molecules.

193. B is correct. Average speed of a gas molecule is inversely proportional to the square root of the molar mass.

194. B is correct. Kinetic energy is proportional to temperature, and speed is proportional to the square root of kinetic energy.

195. A is correct. Average kinetic energy is dependent on temperature only, and thus speed is dependent only on kinetic energy and the mass of the particle, and not on pressure.

196. C is correct. The gas with the greatest molar mass will leak out the slowest and will thus be "left behind."

197. C is correct, Choice *A* can immediately be eliminated because it does not have the correct empirical formula. The speed of diffusion and effusion are both inversely proportional to the square root of the mass. Thus the mass should be inversely proportional to the square of the diffusion rate…1.4^2 is about 2, so the mass should be about twice that of neon. Checking the periodic table gives *C* as the answer.

198. A is correct. This is a fundamental relationship.

199. C is correct. KE = 3/2 kT. Therefore, gases at the same temperature have the same average molecular kinetic energy.

200. A is correct. Since the mass of a molecule of carbon dioxide is more than twice the mass of an atom of neon, the average kinetic energy of the carbon dioxide will be double that of the neon. Since temperature is proportional to average kinetic energy, the temperature will also be double.

201. B is correct. The ratio is found using Graham's law which states that the ratio of the rms velocities of the molecules in a gas at the same temperature is equal to the reciprocal of the ratio of square root of the ratio of the masses. In this case, the molecular weight of HCl is 36 and NH_3 is 17. The square roots of these numbers are approximately 6 and 4 respectively. 4:6 = 1:1.5

202. A is correct. This called diffusion because it is one gas moving through another, in this case the second gas is air.

203. B is correct. The molecules change direction when they collide with other molecules. In this case, the tube contains air, so HCl collides with air molecules. The NH_3 molecules are on the other side of the tube, they also tend to form a precipitate when they collide with HCl, so this is not as good an answer as B. Brownian motion is caused by the outside force of collisions with other molecules.

204. B is correct. Using Grahams law as in question 201, we find the ratio of the average speeds is 4 to 6. so for every 4 cm traveled by HCl, NH_3 will travel 6 cm. Thus they will meet at 4 cm from the left end of the tube and form the precipitate there.

205. D is correct. The sample gas will effuse through the pinhole. It can only reach the pinhole if Stopcock B is open. It will only effuse if there is low pressure behind the pinhole. By opening Stopcock A with the vacuum pump open, pressure is lowered behind the pinhole.

206. C is correct. Gauge 2 measures the pressure in the glass bulb. The gas is effusing from the glass bulb. As the gas effuses, the pressure in the glass bulb decreases, but so does the effusion rate. The effusion rate is represented by the slope of the graph.

207. C is correct. Graham's law says that the molecular weights make a difference. The pressure difference is also required. Under ideal conditions, the molecules are the same size; they have negligible volume.

208. D is correct. The slope represents the rate of effusion. The rate of effusion is a function of molecular mass as per Graham's law. The light the molecule, the great the rate. D shows hydrogen with the greatest slope, then He, then nitrogen.

209. D is correct. Since we're under identical conditions, the most likely reason for non-ideality would be strong intermolecular forces. Choices *A*, *B*, and *C* are all small nonpolar molecules, attracted only by LDF. Choice D exhibits hydrogen bonding, which is a much stronger attraction than LDF.

210. B is correct. As the volume of the container is decreased, the molecules in the container approach each other more frequently, which leads to an increase in the number and strength of intermolecular attractions among the molecules. As these attractions increase, the number and intensity of collisions between the molecules and the wall decrease, thus decreasing the pressure exerted by the gas.

211. B is correct. At lower temperatures, the potential energy due to the intermolecular forces is more noticeable compared to the kinetic energy. The pressure is then reduced because the gas molecules attract each other.

212. C is correct. Real gas molecules have volume making real volume greater than ideal volume. Real gas molecules exert attractive forces on one another slowing them and lowering pressure. The rms velocity of ideal or real gas molecules is directly proportional to the temperature.

213. B is correct. The value of PV/RT for CO_2 at 100 atm is less than one. Real P is lower than ideal P due to attractive intermolecular forces. When you plug a lower P into PV/RT for one mole of gas, you get a number lower than one.

214. A is correct. The value of PV/RT for all gases shown at high pressures is greater than one. Real V is greater than ideal V due to molecular volume. When you plug a greater V into PV/RT for one mole of gas, you get a number greater than one.

215. C is correct. Increased molecular mass and complexity tend to coincide with greater molecular volume and greater intermolecular forces. These cause deviations.

216. A is correct. If the ideal gas law is correct, one mole of gas will result in a value of one for PV/RT when R equals 0.08206 L atm K^{-1} mol^{-1}. According to the graph, this occurs at pressures of 0 atm and 500 atm for CO_2.

217. D is correct. Using the ideal gas law, PV/RT equals 1. In the real methane we can see that PV/RT equals less than one at 200 atm and more than one at 600 atm. So the calculated V must be greater than the real V at 200 atm and less than real V at at 600 atm.

218. B is correct. Liquefaction occurs when molecules are pushed together under high pressure. Since CO_2 deviates from ideal behavior 313 K, it does so at 300 K, a lower temperature, as well. Anyway, since the graph is plotting deviations from ideal behavior, this would not be a good reason to raise the temperature.

219. D is correct. The graph doesn't give information about temperature because the samples are all the same temperature. We can see that at high pressures, PV/RT is greater than one for all samples. A greater than one value for PV/RT indicates a deviation due mainly to volume because real volume is greater than ideal volume.

220. A is correct. Deviations are greatest at low temperatures. No deviation would be a straight line at PV/RT equals one.

221. C is correct. T_3 is the closest to a straight line at $PV/RT = 1$.

222. D is correct. Since $n = PV/RT$, doubling the number of moles, doubles the PV/RT. At 600 atm and 1 mole, T_1 has a PV/RT value of 2.

223. B is correct. T_3 is the highest temperature because it is the closest to ideal. The deviation to T_3 is always positive, which indicates a deviation due to molecular volume.

224. A is correct. *a* reflects how strongly the molecules attract each other. *b* is a measure of the actual volume occupied by just the molecules in a gas. Increasing molecular mass and structural complexity increases attractions and volume.

225. D is correct. See question #217.

226. D is correct. These are all noble gases, so they have the same structural complexity. Thus, *a* and *b* values will tend to increase with molecular mass.

227. A is correct. *a* reflects how strongly the molecules attract each other. *b* is a measure of the actual volume occupied by just the molecules in a gas. Increasing molecular mass and structural complexity increases attractions and volume. This can be seen from the equation.

228. C is correct. Write the balanced reaction: $C_2H_6O + 3O_2 \rightarrow 2CO_2 + 3H_2O$. Since carbon dioxide is produced at twice the rate at which ethanol is consumed, the correct answer is *C*.

229. D is correct. Add the exponents.

230. B is correct. Don't forget $[A] = [A]^1$

231. A is correct. Just use the exponent on [A].

232. A is correct. Reagents that don't affect the rate are said to be of order 0.

233. D is correct. As usual, Roman numeral problems work to your advantage. You probably know item *I* works, so that narrows it down to *A* or *D*. If you then know that either item *II* or item *III* works, you have the answer.

234. B is correct. Remember, rate laws cannot be determined from the balanced equation unless the reaction is known to take place in a single step. We can see from the data, however, that when the concentration is doubled, the rate quadruples. Since four is two squared, this suggests the rate law is second order. If you marked C, you were confusing kinetics with equilibrium.

235. B is correct. Comparing the first two lines, we see that doubling the hydrogen concentration doubles the rate. Since $2^1 = 2$, the reaction is first order in hydrogen. Comparing line 2 to line 4, we see that tripling the NO concentration increases the rate by about nine times. Since $3^2 = 9$, the reaction is second order in NO.

236. D is correct. Any single step reaction has a rate law in which the powers are the number of molecules of a given type which collide. Half a molecule cannot collide!

237. B is correct. Iodide appears only on the right of the reactions.

238. C is correct. 2-methyl propane cation appears first on the right and then on the left, meaning it is produced and then consumed. That makes it an intermediate.

239. A is correct. Water appears only on the left side of the reactions.

240. B is correct. Hydronium appears only on the right side of the reactions.

241. D is correct. Add the individual reaction steps up and cancel anything that appears on the left and right.

242. A is correct. Stay focused on the given information here; we are told that the reaction is first order in the reactant for the first step. That would be the case if the first step were rate-determining. Also, choices *C* and *D* have nothing to do with the order of the reaction.

243. D is correct. NO appears first on the left and then on the right; it is consumed and then regenerated.

244. C is correct. NO_2 appears first on the right and then on the left; it is produced and then consumed.

245. A is correct. O appears only on the left side of the reactions.

246. C is correct. Both intermediates and catalysts cancel when you combine the steps.

247. A is correct. More than one mechanism can generate the same rate law, so scientist A cannot be sure her mechanism is correct based on just the experimental rate law. But if the mechanism is correct, then we know what the rate law should be.

248. D is correct. Distinguish between *rate* and *rate constant*. The rate constant is affected by temperature and by the presence of a catalyst, but reaction rate depends on several other factors.

249. A is correct. The thing you must be careful about here is that the problem asks for what the rate constant depends on, not what the rate itself depends on. The rate constant for a given reaction varies only with temperature.

250. B is correct. A catalyst changes the pathway of a reaction. From the Arrhenius equation, $k = Ae^{-Ea/RT}$, we see that the rate constant depends upon the activation energy E_a. A catalyst lowers the activation energy, and thus changes the rate constant.

251. D is correct. All reactions increase in rate when the temperature is increased, which means choices *A* and *B* can be eliminated. Choice *C* would imply that rate and temperature are directly proportional. Is this true? No. There are a number of ways you can convince yourself of this. For example, you probably know that the rates of many reactions (biological, for example) are exquisitely sensitive to temperature, implying something stronger than a direct proportion (mold which grows rapidly at 310 K might grow extremely slowly at 280 K!) Or, you could know that reactions are generally caused by a very small fraction of the molecules that happen to be moving extremely rapidly; increasing the temperature a little can increase this fraction a lot. Or, of course, you could just know the Arrhenius equation.

252. C is correct. All reactions proceed more rapidly at higher temperatures, yet some reactions are endothermic, others are exothermic. Thus, choices *A* and *B* can be eliminated. And some reactions have lower equilibrium constants at higher temperatures (think of Le Châtelier's principle). So by process of elimination, the only possible answer is *C*.

253. C is correct. If the slower reaction has a much slower reverse reaction, it may eventually dominate. You may perhaps have been tempted by choice *B*. It is a true statement, but it does not invalidate the scientist's claim. The MCAT provides choices like this often; beware of true statements that have no relevance to the question.

254. D is correct. Collisions must have sufficient energy and the correct spatial orientation in order to create a reaction.

255. C is correct. A first order reaction follows the same trend as a half life reaction. The rate is directly proportional to the concentration. On the graph, the slope is the reaction rate. Since the concentration is moving toward zero, the slope should also be moving toward zero. This is curve C.

256. A is correct. The ideal gas law, $PV = nRT$, shows that pressure is directly proportional to moles per unit volume. If we rewrite the given equation as $\log \{[A]_t/[A]_0\} = -kt/2.303$, we see that the units drop out, and only the ratio is important.

257. B is correct. This is just a plug and chug problem. Use the equation: $\log[A]_t = -kt/2.303 + \log[A]_0$. $\log[A]_0 = \log[100] = 2$. $-kt/2.303 = -5\text{x}10^{-5}\text{ x}46{,}000 / 2.3 = -1$. Thus $\log[A]_t = 1$ or $[A]_t = 10$.

258. C is correct. The rate depends upon the specific reaction, not the order of the reaction. A, B, and D are true statements.

259. B is correct. An intermediate is found in a trough. It is a product of an earlier reaction in a multistep reaction that then becomes a reactant in a future reaction.

260. A is correct. The energy of activation is measured from the energy of the reactants to the energy of the activated complex.

261. C is correct. The energy of activation for the reverse reaction is the energy of the products to the energy of the activated complex.

262. B is correct. The energy of activation must be added to begin the reaction, but energy of E_2 is released by the overall reaction.

263. A is correct. A catalyst lowers the energy of activation. The energy of activation is measured from the energy of reactants to the energy of the activated complex.

264. D is correct. As the temperature increases, the collision energies are spread over a larger range with more collisions occurring at higher energies.
265. D is correct. A collision is a necessary, but not sufficient, condition for a reaction.
266. D is correct. A catalyst changes the energy of activation.
267. C is correct. Along with a collision, the molecules must have enough energy and must strike each other with the proper spatial orientation in order for a collision to occur.
268. A is correct. Add the two reactions and cancel NOBr since it appears on both sides.
269. B is correct. The slow step is the rate determining step of a reaction. Any reaction proceed at the rate of the rate determining step.
270. B is correct. The rate of any reaction proceeds at the rate of the slowest step. B gives the rate of the slowest step, Step 2.
271. C is correct. Whenever the fast step proceeds the slow step, the fast step is assumed to reach equilibrium and the equilibrium concentrations are used for the rate law of the slow step.
272. D is correct. Since Step 1 is elementary, we can set the forward and reverse rates equal to each other at equilibrium. The rate laws are $rate_1 = k_1[NO][Br_2]$ and $rate_{-1} = k_{-1}[NOBr_2]$. Setting the rates equal and solving for products over reactants gives $K = k_1/k_{-1}$.
273. B is correct. This is tricky. From the previous question, we know that $K = k_1/k_{-1} = [NOBr_2]/[NO][Br_2]$. Solving for $[NOBr_2]$ we have: $[NOBr_2] = k_1/k_{-1}[NO][Br_2]$. This is the same $[NOBr_2]$ in the rate law of Step 2. Plugging it into the rate law of Step 2 gives: $rate_2 = k_2k_1/k_{-1}[NO][Br_2][NO]$. The rate law for the overall reaction $\{2NO(g) + Br_2(g) \rightarrow 2NOBr(g)\}$ is: $rate_c = k_c[NO]^2[Br_2]$. Since the rate of the Step 2 is equal to the rate of the overall reaction, we set the rates equal and solve. We get $k_c = k_2k_1/k_{-1}$.
274. C is correct. This is the usual case for equilibrium. Incidentally, *D* must be wrong: catalysts, by definition, are regenerated during the course of a reaction, so they should never be used up.
275. B is correct. Because the only difference between the two chambers is the presence of a catalyst, the two chambers will eventually have the same proportion of product to reactant. Initially, however, the chamber with the catalyst will produce product at a greater rate.
276. A is correct. Products over reactants, each raised to the power of its coefficient.
277. C is correct. Pure liquids and solids do not enter into equilibrium expressions.
278. A is correct. Products over reactants; the fact that one is aqueous and one is a gas doesn't matter.
279. D is correct. Pure liquids and solids do not enter into equilibrium expressions.
280. D is correct. Pure liquids and solids do not enter into equilibrium expressions.
281. A is correct. Write $K = [NO]^2[O_2]/[NO_2]^2$ and fill in the given figures.
282. D is correct. This one is a bit tricky, unless you've seen one like it before. All the coefficients of the reaction have been doubled. That means the powers in the equilibrium expression have been doubled, from 1 to 2. But that means all the concentrations have been squared, so the equilibrium constant must have been squared. 530 squared is a lot, so the answer must be *D*. (Don't laugh, of course we could square 530 if we wanted to. But the MCAT is a multiple-choice test. In this case, it's enough to say 530 squared is a lot!)
283. B is correct. Compute the value of the equilibrium expression under these conditions (in other words, the reaction quotient): $5^2/(0.2 \times 0.3)$ is about 350. Since the equilibrium constant is 50, at equilibrium there would be fewer products, and the hydrogen iodide concentration must therefore decrease to reach equilibrium.
284. A is correct. This time the reaction quotient is $6^2/(2 \times 3) = 6$. Since this is less than the equilibrium constant, the amount of product (hydrogen iodide) must increase to reach equilibrium.
285. D is correct. Gases can be written either as concentrations or as partial pressures in equilibrium expressions. The two can be related by the ideal gas law, and thus the numerical values will differ.
286. D is correct. This clearly requires Le Châtelier's principle; we want the equilibrium to shift left. Choices *A* and *B* shift the reaction right. Choice *C* has no effect on equilibrium. Therefore choice *D* must be correct. Notice that it is often effective to use process of elimination on Le Châtelier's principle problems. Perhaps a passage might let you know that this reaction is exothermic, but it is simpler to eliminate the wrong answers.
287. B is correct. Since the reaction is endothermic, you can think of heat as a reactant. Raising the temperature will therefore shift the reaction to the right, thereby producing additional H_2S. This eliminates choices *A* and *C*. Adding more NH_4HS does not affect the equilibrium, since it is a pure solid! Thus, the answer is *B*.
288. C is correct. Use "Roman numeral" problems to your advantage. Item I is in all the answer choices, so you don't even have to think about it—it must be true! Adding an inert gas has no effect on the equilibrium, so option II is not correct. Finally, decreasing the volume of the container shifts the equilibrium toward the side with fewer moles of gas (in this case, toward the right), so III is true.
289. A is correct. Rates always increase when temperature is increased, even when the reaction is exothermic.
290. A is correct. A catalyst increases the rate by lowering the energy of activation. It does not affect the equilibrium.

291. B is correct. Nearly all reaction rates increase with increased temperature regardless of their enthalpy change. The problem is that heat is a product, so the equilibrium would shift to the left according to LeChatelier's principle.

292. C is correct. If the concentrations of two are decreasing and one is increasing, then the two that are decreasing must be the reactants N_2 and H_2. H_2 will decrease three times faster than N_2, so its slope should be three times steeper.

293. C is correct. The concentrations of the reactants and products have stopped changing, but no reactant has been used up, so the reaction has reached equilibrium, has no limiting reagent, and will not run to completion.

294. A is correct. In A, we start with ammonia and run the reaction in reverse. As in the reaction, three times as much hydrogen is formed as nitrogen. B is wrong because more nitrogen is formed than hydrogen. C is wrong because the concentration of nitrogen drops three times faster than that of hydrogen. D is wrong because the concentration of nitrogen rises while the concentration of hydrogen drops.

295. D is correct. A catalyst would cause the equilibrium positions to be reached earlier. This is represented in graph D. Graph C represents no change. Graph A represents the equilibrium positions reached later. Graph B represents a change in the equilibrium concentrations, which does not occur when a catalyst is added.

296. D is correct. By definition.

297. A is correct. By definition.

298. C is correct. By definition.

299. B is correct. Work is being done on the system, so it is not isolated. But no matter moves in or out. (*D* is a nonsense answer.)

300. A is correct. Air, a form of matter, is entering the system.

301. A is correct. This system may be in a kind of dynamic equilibrium, but there is still material entering and leaving it.

302. C is correct. Bell jars are designed so as not to let air or other materials in or out. If it is also insulated, which means that heat cannot get in or out. The fact that the contents do not begin in equilibrium is of no consequence.

303. C is correct. No matter or energy leaks out of the system. Once again, the presence of non-equilibrium processes inside the bunker is irrelevant.

304. A is correct. If matter can leave our universe, it must be open. Incidentally, Stephen Hawking has theorized that black holes also radiate mass *into* our universe.

305. B is correct. Convection through a single styrofoam cup is already effectively zero: convection requires some fluid (liquid or gas) to pass through the surface. But it seems reasonable that a thicker substance reduces conduction. As far as the other choices, *C* is not really desirable for a calorimeter.

306. B is correct. Water removes heat at a greater rate than air does. It removes this heat through direct contact, which suggests the mechanism of conduction. (This is different than the heat removed by water through evaporation, which has to do with the latent heat of vaporization of water.)

307. A is correct. Air and vacuum are the best insulators; storm windows (a window, then air, then another window) also work by utilizing this fact.

308. B is correct. Conduction can be reduced by introducing an insulator; the air or vacuum between the two walls is an effective insulator.

309. A is correct. Since convection conducts heat through fluid motion, the lid prevents fluid (including air) from leaking out.

310. D is correct. A mirror-like surface will reflect radiation, keeping it inside the thermos.

311. B is correct. Hot air is prevented from leaving the system by means of the lid. And no, I didn't make this up—Madam Curie would routinely give this as a question for her hotshot graduate students, almost none of whom came up with this seemingly obvious solution.

312. C is correct. State functions are quantities that, at least in theory, can be measured without knowing anything about the history of the system. Height can be measured with a scale. Cholesterol level can be measured by a blood test. Age cannot be measured in this way; you have to know when the person was born. Admittedly, you might be able to *guess* a person's age by the number of gray hairs, wrinkles around the eyes, or some other measure, but we all know these traits are affected by other factors besides age (sun exposure, for example).

313. A is correct. Work and heat relate to the change in energy of a system when it goes from one state to another. They thus depend on how a system changes between states, and are not themselves state functions.

314. D is correct. Imagine taking a system from state A to state B by one path, and then back from B to A by another. Since *F* is a state function, the value of *F* will be the same the second time the system is at state A as the first time. Thus whatever change is made going from A to B must be undone going from B to A. Since this is true no matter what paths are used, the change in *F* must be path-independent. This is a general property of state functions.

315. C is correct. This is the zero law of thermodynamics: no object can reach absolute zero. If it did, it could no longer radiate heat. We know that A is wrong from practical experience. B is incorrect because a rocket in space moving toward the sun is warmed by radiation. D is wrong because the frequency of radiation does not have to be in the visible spectrum.

316. C is correct. A black body radiator is the perfect radiator. If the emissivity were zero, than *P* would equal zero. Its emissivity should be the maximum possible, which is one.

317. B is correct. A blackbody radiator absorbs and emits perfectly. It's emissivity is one so that *P* is maximized.

318. D is correct. We want to maximize P_n in the second equation. Although delta T remains constant, from the second equation we see that at higher temperatures, P_n is greater. Black objects have a higher emissivity than white objects, which also increases P_n.

319. D is correct. A is a false statement; black absorbs radiation, while white reflects. B is a false statement as well because black robes emit more radiation than white robes. D is a more reasonable explanation than C. Since reflection is away from the robe, most of the energy is reflected away from the wearer of a white robe.

320. C is correct. From the equation $q = mc\Delta T$, we know that the large pot requires more energy than the small pot. Since the pot is losing heat to its surroundings while it is being heated, the most efficient way to heat it is quickly. If we heat it slowly, we give it more time to lose heat to the surroundings.

321. B is correct. White radiates less than black in the dark and reflects more in the sunshine. In the dark, it's cold, so the astronauts don't want to radiate energy; in the sunshine it's warm, so the astronauts want to reflect the energy away.

322. A is correct. Heat flows from hot to cold. If heat is flowing from left to right, the slabs get colder from left to right.

323. D is correct. Like fluid flow or electric current, the rate of heat flow through a slab is constant everywhere.

324. C is correct. This is tricky. From the formula, the temperature difference is directly proportional to the length and inversely proportional to the cross-sectional area. This means that the temperature difference is proportional to the ratio L/A. The passage says that the cross-sections of the slabs are square, so to find the cross-sectional area of a slab, square its height. The ratio L/A for slabs 2, 3, and 4 respectively are: 1/6, 1/2, and 1. Slab 4 has the greatest ratio, and thus the greatest temperature difference.

325. C is correct. Think of this in the extreme case. Imagine a house with thin walls, and a house with thick walls. The outside temperature is 0 °C and the inside of each house is kept at 25 °C. The thin walled house will have the greatest energy bill each month. It loses the most heat per unit of time. Incidentally, since the heat flow through all the slabs is the same while in thermal equilibrium, making slab 3 thinner would increase the heat flow through the other slabs as well.

326. D is correct. This question is similar to question 324. Only now the temperature difference remains constant and k is changing. The slab with the greatest L/A ratio has the greatest k. The L/A ratios for slabs 2, 3, 4, and 5 respectively are 1/6, 1/2, 1, and 2. Slab 5 must have the greatest k.

327. C is correct. The heat transfer is due to molecular collisions within the slab. This is conduction.

328. C is correct. Conduction is due to molecular collisions. When the molecules are far apart, there are fewer collisions and there is greater resistance to conduction. The molecules are farthest apart in a gas.

329. A is correct. Work done *by* a gas is just the negative of the work done *on* the gas. The difference between (or sum of) work and heat is the change in energy. (The whether or not you use difference or sum depends upon the perspective of work and heat.)

330. C is correct. Use the First Law of thermodynamics. One way of writing it is $q_{in} + w_{on} = \Delta E$.

331. B is correct. If each cycle returns to the same state, then the change in energy (a state function) should be zero. But, from the first law of thermodynamics, $\Delta E_{system} = Q_{in} - W_{out}$, or, in this case, $Q_{in} = W_{out} = 60{,}000$ kJ. Some of you who have studied thermodynamics may be bothered by this answer, because you think the second law of thermodynamics demands $W_{out} < Q_{in}$. The problem here is that confusion can sometimes arise between two definitions of Q_{in}. In this problem, we are using it as the net flow of heat into the system. In some situations, this is broken down into a Q_{in} and a Q_{out}, and in that sense, it is true that $W_{out} < Q_{in}$. But that's not the way this problem is set up.

332. C is correct. Work done by a gas is calculated using the following formula: $w = -P\Delta V$. In this case, we calculate it to be $-(50{,}000 \text{ x } ([5-3]) = -100{,}000$ J. But that's the work done *by* the gas. The work done *on* the gas is the negative of this answer, or 100,000 J.

333. D is correct. This is very basic thermodynamics: breaking up water to produce hydrogen, and then using the hydrogen to make water, cancel each other out, which means that there is no net gain of energy (in keeping with the First Law of Thermodynamics). You may wonder, then, what the big deal is about fuel cells, both in the general media, among scientists, and on Wall Street. The answer is that a fuel cell can provide a convenient way to *store* energy. For example, a gas, oil, or nuclear plant produces a lot less pollution per Joule than a car engine. So by running an electric plant and using the energy to split water, and then using the products in a fuel cell in a car, the total amount of pollution can be reduced. Likewise, equipping your house with solar panels normally only gives you power when it's sunny! But a fuel cell could be a convenient way to store the energy for later use.

334. B is correct. The first law is a statement of conservation of energy. Choice *I* is actually a typical *example* of the conservation of energy, with kinetic and potential interchanging indefinitely. So choice *I* is not a violation. This alone yields *B* as the answer. Choice III violates the second law of thermodynamics but not the first.

335. B is correct. The refrigerator described converts 100% of heat into work—a no-no according to the second law.

336. C is correct. Since the gas is kept at constant pressure, we can calculate work with $P\Delta V$. The volume of the gas is decreased, so work is done on the gas. We can also count the squares beneath the function because this gives the change in pressure-volume. There are 16 squares and each square having sides of 50 Pa and 0.5 m^3 is worth 25 joules. 16x25=400. Since we moved from right to left decreasing volume, work is done on the gas.

337. B is correct. There is no volume change, so no work is done. This change occurred completely through heat.

338. B is correct. The volume is increased, so work is done on the surroundings. That's negative work done on the gas. You can not use PV_{final} minus $PV_{initial}$. The squares beneath the path represent the work done. There are 24 squares. Each square is worth 25 J. 24x25=600.

339. A is correct. Like the previous questions, count the squares beneath the graph to find the work done.

340. C is correct. The same path is taken resulting in the work done being reversed on the return trip.

341. B is correct. From A to C work on the gas is negative because the gas is expanding. To find this work measure the area under the curve. From C to D no work is done. From D to A work on the gas is positive because the gas is being compressed. Count the squares under this function to find the work done. Find the difference and that is the net work done on the gas. Or, instead, you can count the squares inside the loop. There are 38 squares times 25 J/square equals 950 joules. The upper path represents more work and is negative so the net work is negative.

342. D is correct. This is just the reverse of the previous problem. Since the same pathway was followed, the work has the opposite sign.

343. A is correct. If volume increases and pressure remains constant, then from PV=nRT we know that temperature must increase.

344. A is correct. The pressure and volume are going up. In order to increase pressure and at the same time increase volume, we must make the molecules move faster by adding energy. We know that work on the gas is negative because the volume is increasing, so energy is leaving the gas via work. Therefore, we must add energy with heat.

345. C is correct. The temperature can be found using $PV=nRT$. However, since P and V increase, so must T. C is the only positive answer.

346. D is correct. The temperature difference depends upon the number of moles, but the amount of heat required to change the temperature of a gas is dependent upon the specific heat of the gas.

347. A is correct. Since there is no heat, q = zero. There is no other energy change, so ΔE equals zero. From $\Delta E = q + w$, we get $w = 0$.

348. A is correct. No work is done and no heat escapes of enters the system, so the average kinetic energy of the molecules doesn't change nor does the temperature.

349. C is correct. Intensive properties do not depend on the amount of the material present. For example, if you take half of a container of water, the density of that half is the same as the density of the full water sample; the density is *not* cut in half. Thus, density is intensive. Specific heat is also intensive, but volume is not. (Volume is *extensive.*)

350. C is correct. Extensive properties are proportional to the amount of material; three times as much mass means three times as much change in the property.

351. B is correct. Intensive properties do not depend on the amount of material.

352. B is correct. Bond energy is internal potential energy; thermal energy is internal kinetic energy.

353. B is correct. For an ideal gas, internal energy is a function of temperature only. Since there are no intermolecular forces in an ideal gas, changing the distance between the molecules while holding temperature constant does not affect the internal energy of an ideal gas. We know that the kinetic energy of the molecules are a function of temperature from K.E. = 3/2 RT. Molecular rotational and vibrational energies are also a function of temperature.

354. C is correct. In the reaction in choice *II*, internal energy merely changes forms, from potential (chemical) to kinetic (thermal). Or perhaps some of the chemical energy is used to do work, in which case the internal energy of the system decreases. In any case, *II* is incorrect, which leaves *C* as the only answer.

355. C is correct. In an exothermic reaction, the bonds formed are stronger than the bonds broken. Note also that answers *B* and *D* are the same thing, and thus cannot be correct.

356. B is correct. This is why any formula involving ΔT can use either Celsius or Kelvin temperature measurements.

357. B is correct. The temperature in this problem is given in terms of change in temperature. Change in temperature is the same when measured in Kelvin or degrees centigrade.

358. C is correct. Thermal kinetic energy is proportional to temperature, but since this is a proportionality, the temperature must be in Kelvins. So the proportion is (273+60)/(273+30), *not* 60/30.

359. B is correct. Filling in $\Delta L/L = \alpha\Delta T$, we get:

$$(1.00018 - 1.0000)/(1.0000) = \alpha\,[10°C - (-5°C)]$$

Rearranging, we get choice *B*. Incidentally, you may be wondering why we didn't have to convert to meters. (Clearly, based on the answer choices we were given, it is unnecessary.) The reason is that the formula involves $\Delta L/L$, and any conversion factor would apply to both the numerator and the denominator, thus canceling out.

360. A is correct. Only pressure changes. The insulation prevents any energy from leaving or entering. The molecules maintain their kinetic energy and thus their temperature.

361. C is correct. Although internal energy still doesn't change, the molecules are attracted to one another. Thus when they are separated, they gain in potential energy. This energy comes from their kinetic energy, lowering their temperature very slightly.

362. B is correct. Heat flows from hot to cold, not necessarily toward less internal energy. For instance, because internal energy is an extensive function, a larger block can have greater internal energy and still be at a lower temperature.

363. D is correct. The equation in the passage says that pressure change is proportional to temperature change. From $PV = nRT$ we know that this is only true if volume stays constant. Also, the passage says that the mercury level on the left must be brought to zero. You can see that this requires the gas volume to be constant. Finally, the reservoir and the right side of the tube must remain at the same level because they are both exposed to atmospheric pressure. You may be able to envision that, since the reservoir is larger than the tube, when it is lowered, more mercury moves into the reservoir lowering both sides of the mercury-filled tube. When it is raised, it holds less mercury, so the mercury in the right and left sides of the tube rises. By raising and lowering the reservoir, the mercury in the left side of the tube is adjusted to the zero mark.

364. B is correct. The mercury in the right side of the tube is lower than the mercury in the left. That means that atmospheric pressure is pushing harder than the pressure of the gas in the bulb. ρgh is a measure of how much lower the gas pressure is.

365. C is correct. The reservoir needs to be lifted so that the mercury in the left side of the tube rises to the zero mark. Since the mercury is below the zero mark, the volume of the gas is too great. From $PV = nRT$ we know that the pressure drops when the volume goes up. The student gets a pressure reading that is too low. From $T = Cp$ we see that a low pressure reading gives a low temperature.

366. A is correct. Using less gas means fewer molecular interactions. Fewer molecular interactions results in behavior more like that of an ideal gas.

367. C is correct. Enthalpy is defined exactly as $U + PV$.

368. B is correct. For an ideal gas, enthalpy is dependent only on temperature. This is because for an ideal gas, internal energy U depends only upon temperature, and because PV equals nRT. Thus enthalpy, $U + PV$, can be expressed as $U + nRT$ for an ideal gas. A function dependent upon temperature only.

369. A is correct. Enthalpy differs from energy in that enthalpy assumes that no work was done by gases. Only in choice *A* does the number of moles of gas change.

370. B is correct. Negative enthalpy changes indicate exothermic reactions. Since the entropy change of the system would be negative, we cannot tell from the information given if the reaction is spontaneous under standard conditions.

371. A is correct. Enthalpy change is an extensive property, so when the coefficients triple, so does the enthalpy change. Admittedly, the kJ/mol notation used in such cases is a bit confusing, but it really means kJ per number of moles shown in the equation!

372. D is correct. Since this is just the reverse of the given reaction, you just change the sign of the enthalpy.

373. B is correct. You must be careful about the definition of heat of formation: the heat required to produce *one* mole of the substance from its constituent elements. Therefore, the reaction we should be considering is:

$$1/2N_2(g) + 3/2H_2(g) \rightarrow NH_3(g)$$

which, compared to the given reaction, is reversed and divided by two. Accordingly, we change the sign on the given $\Delta H°$ and divide by two.

374. B is correct. The heat of formation of water vapor is essentially the heat of combustion of hydrogen—a very exothermic process! But since heat is required to convert liquid into vapor, the heat of formation of water vapor must be exothermic than that of liquid water.

375. D is correct. Don't think of endothermic and exothermic as being related to temperature—think of heat instead.

376. B is correct. This problem requires a bit of consideration, so take your time. (Remember, if you take less time on some problems, you have more time on others.) The heat of formation of $O_2(g)$ is zero, because oxygen is most stable under standard conditions in the diatomic form; therefore we can reject choice *A*. Choice *B* corresponds to the reaction:

$$1/2O_2(g) \rightarrow O(g)$$

Doubling this does indeed give the energy needed to break an oxygen-oxygen double bond, so this should be the answer. Just to be sure, we should look at the remaining choices. Choice *C* can be eliminated because there are no oxygen-oxygen double bonds present in CO_2. While an oxygen-carbon double bond might have a similar bond energy to an oxygen-oxygen double bond, there is no reason to believe that we could obtain a reasonable estimate this way. This is not the correct answer. *D* is also incorrect: double bonds do not, in general, have twice the energy of single bonds.

377. A is correct. Bond energies are generally defined in terms of the energy required to *break* a bond. The energy required to *form* a bond is thus the negative of the bond energy, and has the advantage that you can get heats of reaction by products minus reactants: $2\,[(2(-464)] - [(2\{-436\}) + (-496)] = -488$ kJ/mol. (Some people prefer to keep the signs on the bond energies and take reactants minus products—that's fine too!) As a good double check, you know the formation of water from hydrogen and oxygen is exothermic.

378. A is correct. Bond strengths aren't all that different from each other; they differ by maybe a factor of 2. But hydrogen is *much* lighter than the others, so it's likely to give off the most heat per gram. That's part of the reason that fuel cells, which use hydrogen, are thought of as a possible future source of energy for automobiles.

379. B is correct. Hydrogen, oxygen, nitrogen, and fluorine gases are assigned a heat of formation of zero for their diatomic forms.

380. C is correct. The standard form of oxygen is its diatomic gas. To form atomic oxygen from the diatomic gas, bonds must be broken, which is always an endothermic process.

381. D is correct. A diagram helps here:

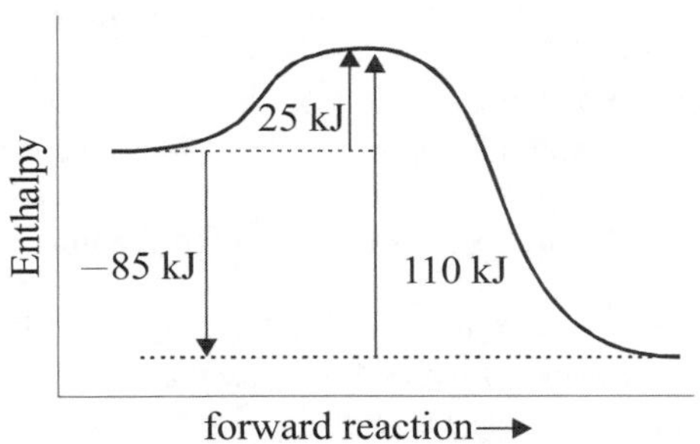

The energy values given in the problem are indicated in the appropriate places on the diagram. Examining the diagram, we can see that the activation energy of the reverse reaction is equal to the sum of the negative of the heat of reaction of the forward reaction and the activation energy of the forward reaction.

382. C is correct. Intermediates are formed and then destroyed. The protons formed in step 1 are consumed in step 2.

383. A is correct. Catalysts are consumed and then regenerated. The bromine consumed in step 1 is regenerated in step 2.

384. A is correct. Formulae can get confusing here; let's try to picture what happens. The experimenter observes a temperature change, and then uses the temperature change to calculate the heat released (or absorbed) in the reaction. But he or she is leaving off some heat capacity when doing the calculation; the experimenter thinks it's easier to change the temperature of the calorimeter than it really is. So the experimenter is underestimating the amount of heat the reaction released or absorbed, which means the calculated value is too low.

385. A is correct. Heat goes from hot to cold. The solution will heat up during an exothermic reaction making the solution hotter than the environment. If the experimenter leaves the lid off, energy escapes from the top and the temperature of the solution doesn't rise as much as it should, so ΔT is low. From $q = mc\Delta T$, we see that the calculated q will be low. That means that the ΔH will be low. The solution will cool down during an endothermic reaction making the solution cooler than the environment. If the experimenter leaves the lid off, energy enters from the top and the temperature of the solution doesn't fall as much as it should, so ΔT is still too low. From $q = mc\Delta T$, we see that the calculated q will be low. That means that the ΔH will be low.

386. B is correct. Choices *A* and *C* misrepresent the role of a catalyst: catalysts do not change the stability of a certain set of products, but they *are* actively involved with reactions (because they change the transition state). As far as choice *D* goes, catalysts certainly do affect the reverse reaction, but that isn't ruled out by the observations; also, what is a "kinetic promoter"? It's a made-up phrase, but it sure sounds like a description of a catalyst!

387. D is correct. $\Delta H = \Delta U + \Delta(PV)$. If P and V don't change much, $\Delta(PV)$ is close to zero. Condensed phases under moderate to low pressures undergo little volume change, so PV work can be ignored. So it's easier to remember that enthalpy change approximates internal energy change in condensed phases not at high pressures.

388. C is correct. The internal energy of a system and the enthalpy of a system cannot be measured. Only changes in internal energy and enthalpy can be measured.

389. A is correct. f stands for formation.

390. B is correct. The superscript indicates standard state, which is 1 bar of pressure (approximately 1 atm) and other things depending upon if the substance is a gas, liquid, solid, solution etc... The superscript usually indicates 25 °C, but, technically, ΔH_f^o can be given at any temperature. 25 °C is not a standard state condition.

391. C is correct. Gases have much higher entropy than other phases.

392. A is correct. Identical masses means the same number of oxygen atoms are present in each case. For choice *A*, the oxygen atoms have maximum freedom—no chemical bonds! Freedom is sometimes a better way to think of entropy than disorder (to a fourth grade teacher, recess may seem like disorder, but to the kids, it's freedom).

393. C is correct. Gases have the highest entropy, so the correct answer choice must have the gas sample listed last; we can thus narrow our choices down to *B* or *C*. Solutions have higher entropy than pure phases of a similar type, since the species are more "mixed up," so the liquid phase will be listed before the aqueous phase in the right answer.

394. C is correct. Entropy has no absolute zero value. By convention, we say a substance at zero Kelvin has zero entropy. This value of zero is arbitrary, not absolute.

395. D is correct. There are many equivalent ways of stating the Second Law, but irreversibility is generally part of a good definition. If a statement implies a process can only go in one direction, it could be a consequence of the Second Law.

396. A is correct. According to the second law of thermodynamics, the entropy gain of the universe must always be positive. The entropy of a system, however, can be negative, as long as the surroundings experience an increase in entropy that exceeds the negative entropy change experienced by the system.

397. D is correct. No way out this time—this would be a blatant violation of the second law.

398. A is correct. Bond formation is always exothermic and always results in an entropy decrease.

399. A is correct. Entropy increases with temperature. Since this is a spontaneous process that does not exchange heat with the environment, the total entropy of the system must increase. Therefore, the gain in entropy by the cold reservoir must be larger than the loss in entropy by the hot reservoir.

400. C is correct. The change in entropy must be positive because, in the isolated system of the two heat reservoirs, heat would spontaneously conduct along the bar from the hotter reservoir to the colder reservoir. To find the exact figure, we use $\Delta S = q/T$ for each reservoir.

401. B is correct. A reaction runs in a direction that increases the entropy of the universe. When it has maximized the entropy of the universe, it has reached equilibrium.

402. C is correct. The blocks became more ordered, so they decreased in entropy. The reaction actually took place, so the entropy of the universe must have been increased. Universal entropy was increased because the man's increase in entropy while stacking the blocks was greater than the decrease of entropy experienced by the blocks. The well ordered nutrient molecules in the man's body were disordered and used for energy to stack the blocks.

403. B is correct. The chemical bond energy of the nutrients within the man went into increasing the potential energy of the blocks and to energy lost as heat. The part that became the potential energy of the blocks retained or increased its potential to do work, but the part that became heat lost much of its potential. Because of the wording of the question, you might have answered C. But B can't be wrong; the heat is definitely at a lower potential. Whereas a portion of the chemical bond energy potential was saved in the potential energy of the blocks. The best answer is B.

404. D is correct. The chemical bond energy was used to stack the blocks and create the heat, so it was the greatest. Any machine is less than 50% efficient, including the body. Recall that aerobic respiration of glucose is 38% efficient. Thus, the heat energy must have been greater than the potential energy of the blocks.

405. D is correct. Entropy is based on probability. The greater the sample space, the greater the entropy. More volume offers more possible positions. Greater numbers offer more objects to fill the positions. Greater temperature means greater movement, so more possible combinations.

406. D is correct. The entropy of a substance always increases with increasing temperature. The equation relates the *change* in entropy at different temperatures.

407. A is correct. A is the most probable system and, therefore, the one with the greatest entropy. At any given moment you are most likely to have two marbles in each side. There are more ways to have one black and one white than to have two whites on the right and two blacks on the left.

408. A is correct. By convention, a substance at zero K has an entropy of zero.

409. B is correct. The change in entropy is the entropy of the products minus the entropy of the reactants. We know that it will be negative because fewer gas moles are found in the products than in the reactants. The reaction is:

$$3H_2 + N_2 \rightarrow 2NH_3$$

Entropy is an extensive property, so we multiply the standard entropies by the number of moles. The entropy of the reaction is –198.3, so we divide by two to get the per mole of ammonia value.

410. B is correct. Entropy increases with temperature for a given substance.

411. C is correct. ΔG only means spontaneous if pressure and temperature remain constant. This is usually true of lab experiments, which are often done at atmospheric pressure and room temperature.

412. B is correct. This formula only requires that temperature be constant. However, if pressure changes, a negative ΔG does not necessarily indicate a spontaneous reaction.

413. B is correct. ΔG indicates the energy made available to do non$P$-V work. This is a useful quantity for things like batteries and living cells where there are no expanding gases.

414. C is correct. Gibbs free energy of the universe is decreased with each reaction, so it does not follow the conservation of energy law.

415. B is correct. The negative of the Gibbs energy is the amount of work that can be done on the surroundings by the reaction. A spontaneous reaction doesn't have to be exothermic, so C and D are wrong. The spontaneity of a reaction doesn't indicate its speed or rate.

416. A is correct. All the thermodynamic functions in this equation refer to the system.

417. D is correct. Negative ΔG indicates a spontaneous reaction, which is always an increase in entropy of the universe.

418. C is correct. If enthalpy change is negative and entropy change is positive, then Gibbs energy change must be negative and the reaction must be spontaneous.

419. A is correct. In this case, the entropy change is inhibiting the reaction. To minimize the importance of entropy, we can lower the temperature.

420. C is correct. Free energies and entropies can be manipulated in the same way as enthalpies. We are given the following data:

$$H_2(g) + \frac{1}{2}O_2(g) \rightarrow H_2O(g) \quad \Delta G° = -229 \text{ kJ/mol}$$
$$H_2(g) + \frac{1}{2}O_2(g) \rightarrow H_2O(l) \quad \Delta G° = -237 \text{ kJ/mol}$$

We want the reaction:

$$H_2O(l) \rightarrow H_2O(g)$$

We can achieve this result by reversing the second reaction and then adding the two reactions: –229 kJ/mol – (–237 kJ/mol) = +8 kJ/mol.

421. A is correct. This is a consequence of the second law. By the way, *C* is backward—the free energy of the universe *decreases*.

422. A is correct. Enthalpy is closely related to energy; when the enthalpy of a system goes down, the enthalpy of the environment generally goes up. This pumps heat into the environment, increasing its entropy as well.

423. B is correct. Consider $\Delta G = \Delta H - T\Delta S$. Since the reaction is exothermic, ΔH is negative. Since the entropy of the system decreases, ΔS is also negative, making $-T\Delta S$ positive. Thus, the two terms are pulling in opposite directions. At high temperatures, the $-T\Delta S$ term wins, ΔG is positive, and the reaction is nonspontaneous. But at low temperatures it will work out the other way, and the reaction will be spontaneous. Of course, the value of the dividing line between "low" and "high" temperatures could be anything, depending on the relative value of ΔH and ΔS.

424. B is correct. In order to be spontaneous at all temperatures, both terms on the right side of $\Delta G = \Delta H - T\Delta S$ must be negative.

425. D is correct. Nonspontaneous reactions can be caused to occur by coupling them to a more spontaneous reaction, so that the total free energy change is negative.

426. A is correct. The standard Gibbs energy for an element in its standard state is zero by convention.

427. C is correct. Since the reaction forms a liquid from gases, we know that the entropy of the system must decrease when the reaction is run. According to $\Delta G = \Delta H - T\Delta S$, such a reaction must be exothermic in order to be spontaneous.

428. C is correct. Notice that the reaction produces a solid precipitate from an aqueous solution, telling us that the entropy of the system must decrease. But the entropy of the universe must be increasing (second law), so the entropy of the surroundings must be increasing.

429. B is correct. Catalysts have no effect on spontaneity; they only affect rate. So if an endothermic reaction is spontaneous with a catalyst, it is spontaneous. This contradicts the scientist's change that all spontaneous reactions were exothermic. If you chose answer *D*, you have to look at the scientist's statement again: all spontaneous reactions are exothermic. That is like saying all elephants have four legs...that doesn't mean that all four-legged animals are elephants!

430. C is correct. First of all, if you marked B, you should rethink your test-taking strategy. Even if you were making a wild guess, there's no reason the reaction with the middle $\Delta G°$ should lead to a maximum yield of nonpolar gases. The formation reactions are of the form:

$$1/2X_2(g) + 1/2Y_2(g) \rightarrow XY(g)$$

Since the more positive $\Delta G°$, the smaller the *K*, NO has the smallest *K*. Since it has the smallest *K*, its equilibrium will lie farthest to the left, so container 3 will contain the greatest proportion of nonpolar gases, in this case nitrogen and oxygen.

431. D is correct. At equilibrium, $\Delta G = 0$, implying that $\Delta H = T\Delta S$. Thus $T = \Delta H/\Delta S = -484{,}000/-89$, which is a bit more than 5000 K. This temperature has to be converted to Celsius, but at such high temperatures, the numbers aren't all that different.

432. A is correct. The reaction shown is just twice the reaction for which the graph is valid. Reading the graph at around 0°C (which is close enough to the 25°C standard temperature for the purposes of this question) gives a result of about –22 kJ/mol. Since free energy change is an extensive property, it must be doubled.

433. A is correct. Consider the equation $\Delta G° = \Delta H° - T\Delta S°$. At 0 K, then, $\Delta G° = \Delta H°$. But since the graph is plotted in degrees Celsius, not kelvins, you must look for the free energy change at −273°C.

434. A is correct. According to $\Delta G° = \Delta H° - T\Delta S°$, the standard entropy change should be the negative slope of a $\Delta G°$ vs. *T* graph. Or, you can "cheat." Combining a gas and a solid into a solid is clearly a process in which the system loses entropy!

435. C is correct. The equilibrium constant equals 1 when the standard free energy change is zero.

436. C is correct. The graph is for the *standard* free energy change. Laboratory conditions are often quite far from standard conditions.

437. D is correct. This time, the experiment was explicitly performed under standard conditions: 1 atmosphere of oxygen gas. Yet the graph shows a negative free energy change at that temperature, which implies that the reaction is spontaneous. Many spontaneous reactions, however, are so slow at certain temperatures that they effectively do not occur. The reaction between oxygen and gasoline is a typical example (you need a spark to get it started).

438. C is correct. Use the equation: $\Delta G° = \Delta H° - T\Delta S°$ and don't forget to convert joules to kilojoules for entropy.

439. D is correct. Solutions can be gases, liquids, or solids.

440. D is correct. "Solution" is really just another name for a homogenous mixture. You might not have realized that alloys are mixtures, rather than true chemical compounds, but choice *III* is certainly a solution, as is choice *I*, which forces you to choose *D*.

441. C is correct. Ideally dilute solutions are so dilute that solute molecules never interact. The mole fraction of the solvent approaches one.

442. B is correct. Ideal solutions obey Raoult's law. The solvent and solute make similar bonds and are similar size and shape.

443. D is correct. You may have heard the phrase "like dissolves like." Students sometimes use this rule to mean all sorts of incorrect things. What it actually means is that polar substances tend to dissolve in polar solvents, and nonpolar substances in nonpolar solvents. Since benzene is nonpolar, we're looking for a nonpolar substance. Silver chloride is ionic, and therefore out. H_2S and SO_2 are both polar; you can verify this by drawing Lewis structures and applying VSEPR theory to find the shape. CO_2 is nonpolar.

444. B is correct. "Like dissolves like" again. Benzene is nonpolar, as is octane. Hydrobromic acid is polar, sodium benzoate is ionic, and sucrose hydrogen bonds.

445. D is correct. Ammonia can form hydrogen bonds. None of the four choices hydrogen bond on their own, but at least SO_2 is polar (see the Lewis structure below):

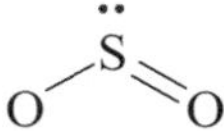

To look at it another way, one of the oxygens in SO_2 has an electron pair free to accept a hydrogen bond from the ammonia.

446. C is correct. Once again, "like dissolves like." Choice *B* is tempting, but has got it backward—the hydrogen bonding in water is stronger than the van der Waals forces in benzene. Choice *C* gives a reason why "like dissolves like" in this case.

447. B is correct. "Like dissolves like" narrows it down to choices *B* or *C*. Again, it's the van der Waals forces in the hydrocarbons that are relatively weak, while the hydrogen bonds between water molecules are strong for intermolecular bonds.

448. D is correct. Choice *B* would certainly encourage solubility, but you should know that the intermolecular bonding in all alkanes is quite similar. In such a case as the one described in the question (an "ideal" solution), the solvation is driven by the entropy of mixing.

449. B is correct. Both choices *B* and *C* give a mechanism for adding mass to the solution, but in case *C* this will not be detected. This is because the mass of a solution is always calculated by taking the total mass of the solution and container and subtracting the mass of the container. If some of the container dissolves, it doesn't matter—it's mass (the *tare*) will still be subtracted from the total.

450. B is correct. A colloid consists of particles larger than those of a solution, but not large enough to settle out due to gravity. C and D are true solutions, not colloids.

451. D is correct. Colloidal particles are too small to be extracted by simple filtration.

452. C is correct. The Tyndall effect results when a beam of light shone through a colloid is dispersed so that the beam becomes visible to the naked eye from the side. In a true solution, the beam shines right through and is invisible from the side.

453. A is correct. In coagulation, heating or adding an electrolyte to a colloid causes the particles to bind together, increasing their mass and allowing them to settle due to gravity.

454. B is correct. The names of polyatomic anions containing oxygen include the suffixes –ite and –ate. –ate is used for the most common oxyanion. –ite is used for the anion with the same charge but one less oxygen. Hypo- means low, indicating the species with one less oxygen than the –ite species.

455. C is correct. The bicarbonate ion is the conjugate base of carbonic acid: H_2CO_3

456. C is correct. The –ate ending indicates the species with more oxygens, but it does not indicate a specific number of oxygens.

457. D is correct. Hydration is described by D.

458. D is correct. When an ion is hydrated, it is surrounded and bonded by water molecules. The average number of water molecules bonding to an ion is that ions hydration number. The hydration number varies, but is often 4 or 6.

459. D is correct. An electrolyte is a substance that conducts electricity in aqueous solution. The fact that it's an electrolyte does not make it a voltage source itself, so no current would flow if it were merely attached to a resistor.

460. B is correct. Strong electrolytes dissociate completely (or almost completely) in water. Choice *I* is in every answer, so we have to take it. Strong acids and bases both dissociate nearly completely, but weak acids and bases dissociate only slightly.

461. D is correct. Potassium oxide contains a metal and nonmetal, which makes it a salt. Salts contain ionic bonds and are strong electrolytes.

462. B is correct. Hydrochloric acid contains two nonmetals, so it must be a molecule. Molecules are covalently bonded. And as you know, hydrochloric acid is also a strong acid, which means, when placed in water, it will break up into in ions. This makes it a strong electrolyte.

463. A is correct. Choice *C* and *D* are salts, and salts are strong electrolytes. Choice *B*, although molecular, is a weak base, and thus a weak electrolyte.

464. C is correct. Acetic acid is a weak electrolyte that is very soluble in water.

465. B is correct. Ammonia is a molecular base making it a weak electrolyte according to the flowchart.

466. C is correct. The chart indicates that molecular bases are weak electrolytes. A is wrong because the word probably implies that some ionic compounds are not electrolytes. B is wrong because some acids are weak electrolytes according to the chart. D is wrong because ionic compounds and acids are not molecular bases and are probably electrolytes.

467. C is correct. From the chart, in order to be a weak electrolyte compound X must be either a molecular base or a weak acid. Since compound X makes $Mg(OH)_2$ more soluble, it is an acid.

468. C is correct. Don't confuse electrolyte strength and solubility. Acetic acid is very soluble but a weak electrolyte. Strong acids are strong electrolytes because they completely dissociate.

469. C is correct. You do not have to memorize solubility guidelines for the MCAT. This is strictly for background knowledge. Alkali metals, nitrates, ammonium salts, and sulfates (with noted exceptions) are soluble.

470. A is correct. This is a common solubility rule, but you do not have to memorize it for the MCAT. It is given here as background information. Halogen compounds are soluble except for salts of silver, mercury, and lead.

471. B is correct. This is a common solubility rule, but you do not have to memorize it for the MCAT. It is given here as background information. Hydroxides are insoluble except for calcium, barium, strontium, and the alkali metals.

472. B is correct. This is a common solubility rule, but you do not have to memorize it for the MCAT. It is given here as background information. Sulfur compounds are insoluble except for calcium barium, strontium, alkali metals, and ammonium salts.

473. C is correct. Molarity is defined as moles per liter of *solution*, not per liter of solvent. The procedure described in *III* would yield 500.0 mL of solvent rather than of solution. Choices *I* and *II* just produce different amounts of the correct solution.

474. B is correct. For molality (lower case *m*!), we need 2 moles of solute per kilogram of *solvent*, so we need to measure out the solvent before the solute is added.

475. C is correct. Mole fraction is not affected by temperature. The mole fraction of a solution is the number of moles of solute divided by the number of moles of solution, and temperature changes do not change the number of moles in a sample. The molarity of a solution is the number of moles of solute per liter of solution; since density changes with temperature, the volume of the solution can change when the temperature changes. The molality of a solution is the number moles of solute per kilogram of solvent; both of these variables are temperature-independent.

476. A is correct. The molality of a solution is the number of moles of solute per kg of solvent. 18.0 grams of sucrose is 0.1 moles. So the molality is 0.1/1.8 *m*.

477. A is correct. For aqueous solutions, molarity and molality are almost (but not exactly) the same, since 1 L of water weighs around 1 kg.

478. A is correct. 18.0 grams of sucrose is 0.1 moles. 1.8 kg of water is 100 moles, since water is 18 g/mol. So the mole fraction is 0.1/(100+0.1).

479. A is correct. The molarity of a solution is the number of moles of solute per liter of solution. 18.0 grams of sucrose is 0.1 moles. The total mass of the solution is 1.44 kg + 0.018 kg = 1.46 kg. With a density of 0.80 g/cm^3, this gives a volume of 1460/.80 $\approx$ 1800 cm^3 = 1.8 L. So the molarity is 0.1/1.8. (Note that the calculations here were a bit of overkill: the density is close to that of water, so a rough estimate would be 0.1/1.4, which is still closest to choice *A*).

480. D is correct. Here's one way to calculate this problem: 1.8 kg of water has a volume of about 1.8 L. 1.8 kg of water is 100 moles, so the molarity is 100/1.8.

481. D is correct. In aqueous solutions, molarity and molality are almost the same.

482. A is correct. Water is 100% water. Mole fractions can never be greater than one.

483. B is correct. Using the periodic table, we see that NaCl has a molar mass of 23 + 35.5 = 58 g/mol. Thus 12 grams is about 0.2 moles. The molarity of a solution is the number of moles of solute per liter of solution, so the molarity of this solution is 0.2/4.0 = 0.05. Since each NaCl dissociates into one chloride ion and a sodium ion, the molarity of the chloride is the same as that of the NaCl solution.

484. B is correct. Using the periodic table, $CaCl_2$ has a molar mass of 40 + 35.5 x 2 = 111 g/mol. Thus 11 grams is 0.1 moles. Molarity is moles of solute per liter of solution, so the molarity is 0.1/4.0 = 0.025. But since each $CaCl_2$ will dissociate into two chloride ions (and a calcium ion), the answer is twice 0.025 *M*, or 0.05 *M*.

485. C is correct. There is no need to actually perform the whole calculation. The concentration of bromide ion in 0.40 *M* NaBr is 0.40 *M*, while the concentration of bromide ion in 0.10 *M* $CaBr_2$ is 0.20 *M* (because there are two bromide ions per $CaBr_2$). When the two are mixed, the concentration of bromide ion must end up between the two. It's like mixing strong coffee and weak coffee; you get in between coffee. There is only one answer between the two initial concentrations, so it must be our answer.

486. C is correct. The molality of a solution is the number of moles of solute per kilogram of solvent. To find the moles of NaOH we need its molecular weight, which, when calculated, is found to be 40. So the number of moles of NaOH is (30 g) (1 mol/40 g) = 0.75 moles. The number of kilograms of solvent is (100 g)(1 kg/1000 g) = 0.1 kg. This gives a molality of 0.75/0.1 = 7.5 *m*.

487. C is correct. Lithium carbonate is Li_2CO_3, since lithium forms +1 ions and carbonate forms −2 ions. Therefore there are two lithium ions per lithium carbonate unit; the correct answer is *C*.

488. D is correct. This is the correct formula for mass percent.

489. D is correct. Although it seems like parts per million should be number of molecules of solute per million molecules of solution, it is based on mass.

490. B is correct. One liter of an aqueous solution weighs approximately one kilogram. milligrams/kilograms = 0.001 $\text{grams}_{\text{solute}}/1000\ \text{grams}_{\text{solution}} = 10^{-6}$ x $(\text{grams}_{\text{solute}}/\text{grams}_{\text{solution}})$ So milligrams of solute per liter of solution represents one millionth of a gram of solute per gram of solution. This is the definition of ppm.

491. C is correct. Parts per million is equal to mg/liter. Parts per billion is one thousand times greater. 0.06 times 1000 = 60.

492. C is correct. Parts per million is equal to mg/liter. 0.0 moles of NaCl is 5.8 grams or 5800 milligrams. Round this to 6000 milligrams and divided by 60 liters. This gives 100 ppm. The closest answer is C.

493. D is correct. The heat of hydration is the enthalpy change when a gaseous solute is dissolved in water. It involves the breaking of water-water bonds, which is endothermic, and the forming of solute-water bonds, which is exothermic. (Since the solute is gaseous, there are no solute-solute bonds.) The net result can be either endo- or exothermic.

494. C is correct. First of all, the concentration is 1 molar, so we are at standard conditions. Since the temperature of the solution drops, we know that the reaction is absorbing heat from its environment, and so we know that the reaction is endothermic. This means $\Delta H°$ is positive. Since the solution is still unsaturated at 1 molar, dissolving more salt at standard conditions must be spontaneous. $\Delta G°$ is therefore negative, and the correct answer is *C*.

495. D is correct. The heat of solution is the enthalpy change (or the energy absorbed as heat at constant pressure) when a solution is formed. In order for the solution to be formed, solvent-solvent bonds must be broken, solute-solute bonds must be broken, and solute-solvent bonds are formed. The breaking of bonds absorbs energy, while the formation of bonds releases energy. If the heat of solution is negative, then energy is released.

496. D is correct. If the bonds formed have lower energy than the bonds broken, the reaction is endothermic. Therefore, using $\Delta G = \Delta H - T\Delta S$, the entropy of the system must increase if the reaction is to be spontaneous.

497. B is correct. The formation of bonds releases energy. Heat of solution is the sum of the enthalpy changes for the breaking and formation of bonds. During solution formation, solvent-solvent bonds are broken; solute-solute bonds are broken; and solute-solvent bonds are formed. Thus, if the heat of solution is negative, the solute-solvent bonds are stronger than the solvent-solvent bonds and solute-solute bonds.

498. B is correct. In an ideal solution, the bonds formed are similar to the bonds broken resulting in a zero heat of solution.

499. B is correct. We need to reverse the reaction of the formation of NaCl from its gaseous ions, and change the sign of its enthalpy change. We add this to the reaction for the heat of hydration.

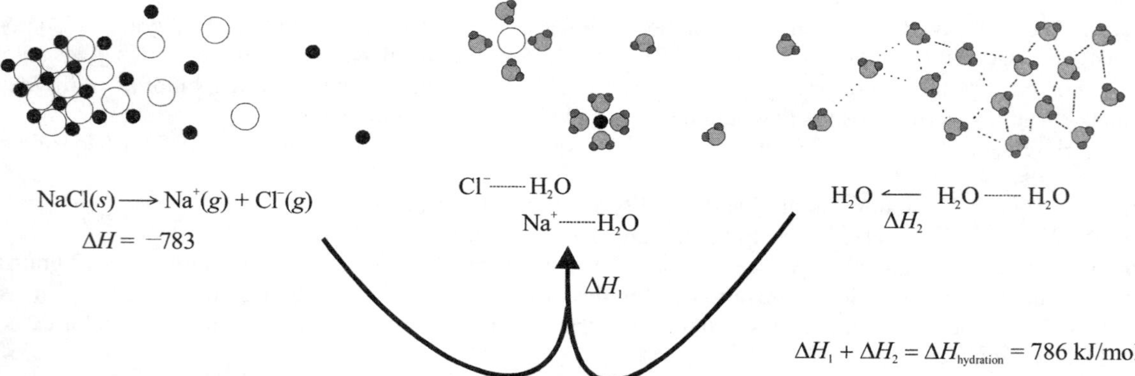

500. C is correct. NaCl is very soluble in water due to the large increase in entropy. For practical purposes, the dissolution of NaCl in water is spontaneous. At most temperatures, sodium chloride will dissolve spontaneously in water. $\Delta G = \Delta H - T\Delta S$. However, since enthalpy change is positive, there is some temperature dependence involved. In addition, choices A and B indicate that spontaneity depends upon enthalpy change only. This is incorrect. Although the change in enthalpy changes slightly with temperature, for MCAT, you should assume that the change in enthalpy for a reaction remains constant with temperature change. This makes the best answer C.

501. D is correct. The heat of hydration is the sum of the endothermic breaking of the hydrogen bonds of between water molecules and the exothermic formation of the bonds between water molecules and ions. Since the heat of hydration is negative, the energy released by the formation of water-ion bonds must have a greater magnitude than the energy absorbed by the breaking of hydrogen bonds. There bond energy is a measure of their strength.

502. A is correct. It is a general trend that the solubility of salts increases with temperature. There are, of course, exceptions to this rule. Adding water will increase the amount of NaCl that will dissolve, but it will not increase the concentration.

503. C is correct. The heat of solution is positive and the heat of hydration is negative. The heat of hydration must come after the dissociation of the NaCl ions.

504. B is correct. When the oil molecules are present in the water, the water molecules orient to form a solvent cage around the oil. This reduces the entropy of the water and thus makes the interaction unfavorable.

505. D is correct. This is how vapor pressure is normally defined.

506. B is correct. Think of this as an equilibrium: the rate at which the solid turns into vapor must be greater than the rate at which the vapor turns into solid. Consideration of the kinetics suggests that the rate at which the vapor turns into a solid should be proportional to the partial pressure of the vapor.

507. C is correct. After reading the explanation to the previous question, you might think the correct answer is *B*. But *B* is the condition for *evaporation*. *Boiling* is a different process. Most people think that, at one atmosphere pressure, water is a solid below 0°C, a liquid between 0°C and 100°C, and a vapor above 100°C. But this is not quite correct. At all temperatures, at equilibrium, some proportion of the water will be in vapor form. The boiling point is the temperature above which liquid can no longer exist (assuming equilibrium), but vapor can certainly exist below the boiling point.

508. B is correct. The normal boiling point of a liquid occurs when the vapor pressure of the liquid reaches one atmosphere.

509. C is correct. Vapor pressure always increases with increasing temperature.

510. C is correct. Since water boils when the vapor pressure equals the atmospheric pressure, the atmospheric pressure must be equal to the vapor pressure of water at 97°C. Water has a vapor pressure of 760 torr at 100°C. Vapor pressure decreases with decreasing temperature. Thus the vapor pressure of water at 97°C, and thus the atmospheric pressure in this problem, is less than 760 torr. Condensation occurs when partial pressure is greater than or equal to vapor pressure. Since, in the problem, water condenses on a surface at 13°C, the partial pressure of water vapor in the air must be at least as great as the vapor pressure of water at 13°C.

511. D is correct. How can A and B both be true? A is at a given pressure and B is at a given temperature. In other words, A says that if we hold pressure constant, the vapor pressure of a liquid is greater than a solid. That's because the liquid turns into a solid when the vapor pressure of that liquid drops to the vapor pressure of the solid. B says that if we compare the vapor pressures below the melting point, the solid has a higher vapor pressure. We start with a liquid above the melting point. Here its vapor pressure is higher than the solid's. As we decrease the temperature, both vapor pressures lower. However, the liquid's lowers faster. At some low temperature, the two are equal. If we lower the temperature even further, and lower the pressure to keep the liquid from freezing, the vapor pressure of the liquid is lower than the vapor pressure of the solid at the same temperature.

512. B is correct. This is LeChatelier's principle. By the way, the term steam is reserved for water vapor over 100 °C, so C and D are wrong.

513. B is correct. Pure water has a vapor pressure of 760 torr at 100°C, the temperature at which it boils. A 2 *m* sodium chloride solution is 2 *m* in Na^+ and 2 *m* in Cl^-, for a total of 4 *m* in solute. For convenience, consider 1 L of solution. There are 4 moles of particles from the salt, plus about 56 moles of water (this is a handy number to remember). So the mole fraction of water is 56/(56+4), or more than 0.9. Raoult's Law states that the vapor pressure of one component of a solution is equal to the vapor pressure of the component when pure multiplied by its mole fraction. So the vapor pressure here is about 0.9 x 760, or *B*. Alternatively, save yourself the calculation and think about what a reasonable number might be in this situation: the vapor pressure must be less than 760 torr, but a 2 *m* NaCl solution won't have a radically lower vapor pressure than pure water.

514. A is correct. At higher temperatures, vapor pressures are higher; also, nonvolatile solutes (such as salts) lower vapor pressure, so choice *A* wins on both counts. If choice *C* or *D* had specified a temperature greater than 30°C, we would not be able to determine the correct answer without a table, graph, or equation giving vapor pressure as a function of temperature.

515. C is correct. Skip the long Raoult's law calculation: the vapor pressure of an ideal solution must be intermediate between the vapor pressure of its components, and only one answer satisfies this requirement.

516. D is correct. Vapor pressure goes up with temperature. Steam has a vapor pressure above 760 torr, and water has vapor pressure below 760 torr, so D is correct. Solids have lower vapor pressure than liquid, so A is wrong.

517. D is correct. D is false because the atmospheric pressure equals the vapor pressure at the boiling point. A is true by definition. B is true by observation of the graph. C is true because the liquid would be a solid at temperatures below the melting point unless the pressure were reduced, and the solid would be a liquid at temperatures above the melting point unless the pressure were increased. Supercooled water is used for measurements of liquid vapor pressure below the melting point and extrapolation techniques are used for the solid above the melting point.

518. A is correct. High altitudes have lower atmospheric pressure. Water boils when its vapor pressure equals its atmospheric pressure, so water boils at a lower temperature and vapor pressure at high altitudes. It's not the boiling that cooks the egg, it's the temperature. That means you can cook it faster in the warmer water of low altitudes.

519. A is correct. The scientist rearranges Raoult's law to solve for mole fraction: $X_a = P_a/P_a°$. The actual solution vapor pressure is lower than the Raoult's law prediction. Therefore, in order to use Raoult's law, the scientist should have substituted a *higher* value for P_a, and the correct value of X_a is also higher. Therefore, the estimate is too low.

520. B is correct. The easiest way to answer this question is to know that adding a nonvolatile solute lowers the vapor pressure, and to see realize that the vapor pressure of pure water at 100 °C is 760 torr. Thus the answer is A or B. The mole fraction of water is much greater than the mole fraction of NaCl, so the vapor pressure could not be cut in half and A must be wrong. The hard way to answer the question is as follows: There are 55.5 moles of water in one liter. 58 grams of NaCl represents one mole. The mole fraction of water in this solution is 55.5/(55.5 + 1) = 0.98. Use Raoult's law ($P_v = \chi_w P_w$) and multiply the mole fraction by 760, the vapor pressure of water at 100 °C and we get 747 torr.

521. D is correct. A is incorrect, because if both substances are volatile, making a solution could raise the vapor pressure. B is false. C is wrong because you can have deviations to Raoult's law causing a lower of vapor pressure even when adding a substance with a higher vapor pressure. A is an OK answer, C is a better answer, but D is the best answer. MCAT says to choose the *best* answer.

522. D is correct. The pure water has a greater vapor pressure than the salt solution. This means that molecules of water are escaping faster from the pure water than from the salt solution. This will continue to be the case until the two glasses have the same mole fraction of solute and thus the same vapor pressure. Since glass X has a mole fraction of zero solute, glass Y will never be that dilute. All the solvent (water) will move into glass Y.

523. B is correct. This is the definition of a nonvolatile solute.

524. C is correct. Vapor pressure is independent of atmospheric pressure.

525. C is correct. The rate of evaporation is dictated by the temperature, so it remains constant. As molecules fill the space above the liquid, the rate of condensation (molecules reentering the solution) increases until the rates are equal.

526. A is correct. Pressure is created by molecules. The only source of molecules in space Q is the fluid surface. Once an equilibrium is reached, the pressure is the vapor pressure of unknown fluid.

527. C is correct. The surrounding atmospheric pressure holds up the column of fluid in the barometer. Thus, the surrounding fluid pressure equals ρgh of the fluid plus the vapor pressure of the fluid. The greater vapor pressure of water would push down on the column. Point D would represent a column height of zero, and the vapor pressure of water is not equal to one atmosphere so *h* would have some positive value. The value of *h* can be calculated by setting 105 Pa equal to ρgh minus the vapor pressure.

528. B is correct. If the fluid level in the column were equal to the outside surface level, the vapor pressure would be pushing on the fluid with the same pressure as atmospheric pressure.

529. C is correct. "Oh," you thought to yourself when you read this problem, "this is going to be an easy one. Like dissolves like; I'll just look for a nonpolar thing, and it won't be soluble in water." Good thought, but all the answers here are salts, so there must be another method. One way is to have memorized that the salts of alkaline metals, ammonium, and halogens are always quite soluble in water. Another is to realize that MgO involves a +2 and –2 ion; the higher the charges on the ions in a salt, the more difficult it is to separate them when dissolving the salt in water.

530. B is correct. Ions not involved in the reaction are called spectator ions.

531. D is correct. Insolubility is a relative concept. Typically, compounds with solubilities of less than 0.01 mol/L are considered insoluble. With regards to answer choice B, on the MCAT you can assume the solvent is water unless otherwise stated.

532. B is correct. You may recall that the hydroxide of Ba is considered soluble while the carbonates are insoluble except for alkali metal and ammonium compounds. The sulfides are insoluble except for alkali metal, ammonium ion, and Ba^{2+}, Ca^{2+}, and Sr^{2+} compounds. Or if you remember the carbonate rule but not the sulfide rule, you might notice that the K_{sp} for $NiCO_3$ is larger than for FeS and thus must be FeS is less soluble. Barium hydroxide dissociates into three particles, so the equation will be $K_{sp} = [Ba^{2+}][OH^-]^2. = (x)(2x)^2$. For barium hydroxide x is greater than 0.01. $NiCO_3$ and FeS each dissociate into two particles, so their solubility is given by $K_{sp} = x^2$. x for these compounds is less than 0.01.

533. A is correct. Ion pairing removes ions from the equilibrium expression shifting the equilibrium to the right. B is wrong because we don't consider whether or not the reaction is exothermic when finding solubility with K_{sp}. C was already taken into account by the equilibrium expression. D is already taken into account by the equilibrium expression and a basic solution would shift the equilibrium to the left anyway.

534. B is correct. Typically in this type of question, the compound with the lowest K_{sp} is the least soluble. Although $FeCO_3$ has a larger *Ksp* than Ag_2CO_3, Ag_2CO_3 breaks into three particles which makes it more soluble for a given K_{sp} value.

535. A is correct. Silver nitrate is soluble. The silver ions dramatically decrease the solubility of the silver carbonate due to the common ion effect. Thus since the remaining compounds all dissociate into the same number of particles, we look for the compound with the greatest K_{sp}.

536. A is correct. Acidic solutions have fewer hydroxide ions and increase solubility of calcium hydroxide based on the common ion effect.

537. D is correct. The solubility product does not change with changing ion concentration. It only changes with temperature.

538. C is correct. Nickel ions form insoluble compounds with each anion. Compare the K_{sp} values to decide which ones will precipitate first. Don't do too much work here. The solubility is the square root of the K_{sp} for compounds that dissociate into two particles, and the cube root (of one fourth the K_{sp}) for compounds that dissociate into three particles. For a quick estimate, divide the exponent on the K_{sp} by 2 for two particle compounds and by 3 for three particle compounds. The approximate solubilities are $10^{-3.5}$, $10^{-5.3}$, $10^{-10.5}$

539. C is correct. In this particular case, the solubility in water is the square root of the K_{sp}.

540. B is correct. The solubility will decrease due to the hydroxide ions in solution. A is the solubility in neutral water, so A is wrong. To calculate the value we have: $K_{sp} = [Ca^{2+}][OH^-]^2. = (x)(2x+0.1)^2$. The 0.1 represents the hydroxide ions contributed by 0.1 M KOH. x is likely to be much smaller than 0.1, so we ignore it for now and verify this assumption later. Solving for x, we get $x = 1.3 \times 10^{-4}$. We see that x is, in fact, much smaller than 0.1.

541. C is correct. Ion pairing and hydrolysis reactions take ions out of solution pulling the equilibrium to the right and dissolving more ions.

542. B is correct. Recall that alkali metal salts are generally quite soluble in water, so the answer is not likely to contain sodium.

543. C is correct. Here is he reaction: $Ca_3(PO_4)_2(s) \leftrightarrow 3Ca^{2+}(aq) + 2PO_4^{3-}(aq)$. When we write an equilibrium expression, the coefficients become exponents, the products are placed over the reactants, and pure solids and liquids are left out.

544. C is correct. One common definition of a saturated solution is one that is in equilibrium with its precipitate. Another is that if the ion-product is equal to the K_{sp}, the solution is saturated. These two definitions are not the same. MCAT will not make you choose between two working definitions, so don't worry if you missed this question. The question is here to make you aware of the two definitions, and make you think about the difference. In the first definition, if you heat the solution, it remains saturated, while by the second definition, the solution may become unsaturated because it is not in equilibrium with its precipitate.

545. A is correct. Look at the overall reaction at the bottom. LeChatelier's principle tells us that adding HF pushes the reaction to the left creating a precipitate.

546. C is correct. Again, looking at the overall reaction, the chloride ions are spectator ions, leaving protons to push the reaction to the right. D is wrong because there is no leftward shift caused by HCl.

547. C is correct. When the acid was added, the dissociation equilibrium shifted to the right allowing room for more calcium.

548. A is correct. Here is the reaction for the dissolution of sucrose in water: sucrose(s) $\rightarrow$ sucrose(aq). Since solids are not included in equilibrium expressions, the correct answer is *A*.

549. B is correct. The reaction is $O_2(g) \leftrightarrow O_2(aq)$. When writing an equilibrium expression, the coefficients become powers, the products are placed over the reactants, and pure solids and liquids are left out, but gases are included.

550. D is correct. Changing the temperature changes the solubility constant. Choice *B* describes a *supersaturated* solution, and choice *C* describes a *saturated* solution.

551. A is correct. The dissolution of $PbCl_2$ is given by the reaction: $PbCl_2(s) \rightarrow Pb^{2+}(aq) + 2Cl^-(aq)$, thus giving $K_{sp}=[Pb^{2+}][Cl^-]^2$. In a 0.001 M $PbCl_2$ solution $[Pb^{2+}] = 0.001$ and $[Cl^-] = 0.002$, so $Q = (0.001)(0.002)^2 = 4 \times 10^{-9}$. Since $Q < K$, the solution is unsaturated.

552. A is correct. Both sodium and nitrate ions form highly soluble salts, so choices *B* and *D* are out. The dissolution of $PbCl_2$ is given by the following reaction: $PbCl_2(s) \rightarrow Pb^{2+}(aq) + 2Cl^-(aq)$, giving $K_{sp}=[Pb^{2+}][Cl^-]^2$. The final volume of the mixture is 1000 mL, so there will be some dilution. $[Pb^{2+}] = (0.01)(900/1000) = 0.009$, while $[Cl^-] = (0.1)(100/1000) = 0.01$. So Q is less than $(0.009)(0.01)^2 = 9 \times 10^{-7}$, which is less than K_{sp}. Therefore, the solution is unsaturated.

553. A is correct. Despite what the answer choices might lead you to believe, this problem has nothing to do with acids and bases, since NaCl is neutral. Le Châtelier's principle (or, if you like, the common-ion effect) can be used, however:

$$NaCl(s) \rightarrow Na^+(aq) + Cl^-(aq)$$

Because hydrochloric acid will add chloride to the solution, it will drive the reaction to the left and reduce the solubility of sodium chloride.

554. B is correct. For a given reaction, K_{sp} depends *only* on temperature. Solubility, on the other hand, is also affected by the common ion effect.

555. A is correct. See the explanation to the previous question.

556. C is correct. The least soluble compound will precipitate first if the concentrations are equal, as they are in this problem. There is one caveat, however: it is wrong to compare K_{sp}'s of salts that break up into different numbers of particles, since the "units" are different. For example, we could not compare the K_{sp} values for $CaCO_3$ and Na_2CO_3, since they dissociate into two and three ions, respectively. In the case we find in the problem, all the salts break up into one cation and one anion, so we may compare the K_{sp}'s.

557. D is correct. Don't try to do all the stoichiometry; look instead for the differences among the choices. The only differences involve the 20.0 and the 6.0; we need to decide whether they should be in the numerator or the denominator. The 20.0 must be the volume of silver nitrate. It stands to reason that the greater the volume of the solution, the more sodium chloride has to be added. Therefore 20.0 should be in the numerator, which narrows our choices to *A* or *D*. On the other hand, the 6.0 *M* is the concentration of silver nitrate. The precipitate should form more readily in a more concentrated solution; in other words, the higher the concentration of silver already present, the less sodium chloride must be added in order to bring about precipitation. Thus, we should look for 6.0 in the denominator. We had already narrowed the choices down to *A* or *D*, and now know the correct answer is *D*.

558. C is correct. First, check your vocabulary: iodine is I_2, iodide is I^-. The reaction of iodine with iodide continually removes iodine from solution. (The reaction happens to be $I_2 + I^- \rightarrow I_3^-$.) By Le Chatelier's principle, the equilibrium for the reaction $I_2(s) \leftrightarrow I_2(aq)$ shifts to the right. A few brief words on the wrong answers. For choice *A*, there are two problems: first, iodine and iodide are different species; second, the common ion effect suppresses solubility, rather than enhancing it. Choice *B* sounds plausible, except that iodide should be a reducing agent! (Indeed, the reaction alluded to in choice *C* can be considered a reduction.) Choice *D* confuses iodine and iodide.

559. D is correct. Adding sodium chloride will certainly decrease the solubility via the common ion effect, so II is correct. Evaporating water will increase the concentration of the ion, so III is also correct. The only answer choice containing II and III is D. (It would have been a reasonable bet anyway, since the solubility of most salts decreases with decreasing temperature, but that is not always the case.)

560. B is correct. Le Chatelier's principle gives us this answer.

561. B is correct. Gases become less soluble in water as the water temperature increases; at higher temperatures, the increased entropy of the gas is more of a factor.

562. D is correct. I. Gases are less soluble at low pressures. II. Shaking causes the bubbles to coalesce. III. Gases are less soluble at high temperatures. IV. Salt crystals nucleate the gas bubbles and cause them to coalesce.

563. C is correct. The cold waters of the Arctic contain the most oxygen, and can support the most life. Think about where you find large sea mammals such as whales and walrus.

564. C is correct. From the equation, the partial pressure is directly proportional to the solubility.

565. D is correct. If the gas reacts chemically, more gas can dissolve in the solvent. In addition, pressures should be low for Henry's law to be accurate.

566. D is correct. *P* for oxygen is 4 times less and *k* is approximately 2 times less. Both are proportional to the mole fraction in water, so 4x2=8 and there is 8 times less oxygen than nitrogen in water according to Henry's law.

567. A is correct. Before you begin calculations, do a little process of elimination. D is an impossible value for mole fraction, which can't be greater than one. C would indicate that water is 88% oxygen, and B would indicate that water is 8.8% oxygen. Both are ridiculous. So A must be the answer. Use Henry's law to find the answer. The partial pressure of oxygen is 10^5 Pa times 0.2. Changing this to kPa leaves 2×10^1 kPa. Multiply this by Henry's law constant as in the formula and we have 8.8×10^{-3}.

568. B is correct. The dotted lines follow Henry's law. The real gases in the graph show deviations at high pressures.

569. B is correct. Molecules of a solid vibrate in place, even at zero K.

570. C is correct. The molecules in a liquid remain close, but move past each other breaking and forming bonds.

571. D is correct. The molecules in a gas don't bond to each other. In an ideal gas they experience no attractive forces whatsoever.

572. B is correct. Read all the answer choices! When carbon dioxide is dissolved in water, it does *not* liquefy.

573. D is correct. The important thing to know here is that temperature, pressure and number of moles is enough to define a phase of a given substance. This is not crucial information for MCAT takers, but it helps understanding.

574. C is correct. If you marked *D*, think of the more familiar example of an aqueous solution: we know it is *one* phase. If you marked *A* or *B*, then realize that a phase must be *homogenous*; if you can distinguish substances on a macroscopic scale, there is more than one phase present.

575. B is correct. The description suggests chocolate is an *amorphous* solid. Amorphous solids do not have well-defined melting points.

576. C is correct. Ice, water, and water vapor.

577. B is correct. There is no hydrogen bonding in hydrocarbons; choice *A* is eliminated. Saturated hydrocarbons do not form dimers; knock out choice *C*. As for choice *D*: why would energy be needed for a hydrocarbon to freeze? Usually, energy is released when a substance freezes.

578. C is correct. The ether and water will separate into two liquid phases due to the polarity difference, but their vapors will mix to form one gas phase. Gases form homogeneous mixtures with each other regardless of their identities. Gases nearly always form one phase.

579. B is correct. Heat capacity is a measure of how hard it is to change the temperature of a substance. Since block X changes temperature more easily, it must have the lower specific heat.

580. B is correct. A is a possible answer, but the best answer here is clearly B. "Specific" in thermodynamics indicates per unit mass. Specific heat is the energy necessary to change the temperature of 1 g of a substance by 1 K. Heat capacity is a more general term that can be used when referring to complex systems such as an entire calorimeter including the stirring apparatus, the thermometer, etc…

581. B is correct. A greater proportion of heat energy is absorbed by the vibration of the hydrogen-oxygen bonds of water than by the bonds of most molecules. *A* is wrong because temperature is proportional to the square of the velocity: KE = 3/2 kT. *C* is wrong because bond breakage would lead to a phase change; it is not a significant factor in determining the heat capacity of a phase. *D* is wrong because bond formation releases energy.

582. B is correct. Use $q = mc\Delta T$: 10 (0.24) (25 – 20) = 12.

583. C is correct. Use $q = mc\Delta T$: 20 = 10(0.25)ΔT, so ΔT = 8. Since the temperature starts at 12°C, the final temperature is 20°C.

584. D is correct. Heat capacity is equal to *mc*, and the specific heat of water is 1cal/g K, so 15 grams of water has a heat capacity of 15 x 1 = 15 cal/K.

585. B is correct. Molar heat capacity is the heat capacity per mole of the substance. A mole of water weighs 18.0 g, and the specific heat of water is 1.0 cal/g K, so the molar heat of water is 18.0 cal/mol K.

586. C is correct. A high heat capacity means that water gain or lose energy with smaller changes in temperature. This corresponds to I and III only.

587. C is correct. Don't get lost in equations. If it takes 111 J/mol to raise the temperature of ethanol 1 K, then 3 moles must require 3 x 111 J. But we want a 5 K temperature change, so we need 5 x 3 x 111 = 1700 J.

588. B is correct. There are several common errors addressed by this question. A more accurate equation would be $w + q = mc\Delta T$ since work is energy transfer as well. Work is typically left out by chemistry books because systems in chemistry are often at rest with no *PV* work being done. *m* is mass, not molality. ΔT is the change in the temperature not the temperature of the substance, and ΔT may be in kelvins or degrees Celsius because a change in temperature is the same value for either.

589. C is correct. The temperature of water at 100 °C does not increase when heat is added, regardless of how much heat. Instead it turns to steam. The specific heat of a substance at its boiling point is not something you should memorize for the MCAT; it is something that you should understand for the MCAT. Notice how the answer choices were accompanied by explanation to guide you to the correct answer.

590. D is correct. I. The gas does work on the surroundings, lowering its energy and thus its temperature. II. You should know the relationship between temperature and the average kinetic energy of fluid molecules. K.E. = 3/2 RT. III. A force is exerted on the piston over the distance through which is moves. $W = Fd$.

591. C is correct. The molecules will have more kinetic energy, so temperature is increased. The volume will remain the same. Since pressure is kinetic energy per unit volume, more kinetic energy with the same volume means greater pressure.

592. B is correct. A greater heat capacity would mean less change in temperature for the same amount of heat. From PV=nRT, we see that a smaller change in temperature corresponds to a smaller change in pressure.

593. C is correct. From a thermodynamics view point: According to the ideal gas law, if the volume increases and the pressure is constant, the temperature must increase. But since the gas is expanding against an external pressure, energy must also be expended to do work. You can also view this from a kinetics viewpoint: The molecules collide with the piston and the piston gives way taking kinetic energy from the molecules. So some of the energy goes into work on the surroundings and some goes into kinetic molecular energy of the molecules.

594. A is correct. Heat capacity is a substance's ability to absorb energy with as little change in temperature as possible. If a gas is allowed to expand doing work on its surrounding, it is transferring away some of the added heat energy and thus can absorb more heat with less change in temperature.

595. D is correct. A and B are wrong because the amount of force is a function of the pressure not a function of the structure of the solid. At short distances, electrostatic forces between molecules are subject to large fluctuations as they change with distance. Look at Coulomb's law: $F = kqq/r^2$.

596. C is correct. Specific heat is an intensive property. It doesn't change with the amount of substance.

597. A is correct. Heat capacity is an extensive property. It increases when more substance is available to absorb the energy.

598. A is correct. Heating a substance cannot, in itself, cause that substance to cool as would be indicated by a negative heat capacity.

599. A is correct. The gold bar has the lowest heat capacity and, assuming that each object is absorbing energy at the same rate, the gold bar will change its temperature the fastest.

600. B is correct. Water has the highest heat capacity, so it can absorb the most energy with the smallest temperature change.

601. B is correct. The molecular structure of methane allows it to absorb energy in the form of bond vibration, molecular rotation, things that the monatomic argon cannot due. C is wrong because methane molecules move faster, but specific heat is related to energy, not speed, and methane molecules and argon molecules have the same average kinetic energy at a given temperature. D can't be correct because specific heat is an intensive property and argon isn't heavier than methane. Anyway, what exactly does heavier mean? Greater density? Greater molecular weight?

602. D is correct. The larger, more complicated molecule is likely to be able to absorb more energy vibrationally, rotationally, translationally, and intermolecularly keeping its temperature down.

603. A is correct. Energy was lost by Liquid B and gained by Liquid A. Since the system was insulated, no heat was transferred to the surroundings, so energy lost by Liquid B went to Liquid A. Q was the same for both liquids. The temperature change was the same. From $Q = mc\Delta T$ we see that c is inversely related to m. Negative specific heats are impossible.

604. A is correct. Reactions in coffee cup calorimeters take place at constant pressure. At constant pressure, the energy transfer (q) is approximately equal to the enthalpy change.

605. B is correct. The pressure in the solution is 1 atm plus ρgh. 1 atm of pressure is equal to an h of about 10 meters (or 30 feet) in aqueous solution, so the ρgh term is insignificant.

606. A is correct. The specific heat of a dilute aqueous solution of sodium chloride can be approximated by water.

607. B is correct. The stirring mechanism creates a homogeneous mixture and uniform distribution of temperature, both part of the definition of a single phase.

608. C is correct. Enthalpy change in an extensive process. Twice as much solution should result in twice the enthalpy change.

609. B is correct. The molar heat of solution is an intensive property. It doesn't change with amount.

610. B is correct. Don't bother with a detailed calculation! There's more of the 25°C water than 80°C water, so the result must be closer to 25°C (and between the two, of course).

611. D is correct. If the two compounds had simply mixed and not reacted, the temperature would have been between the two starting temperatures. But in this case the result is hotter than either of the starting solutions, so there must have been an exothermic reaction.

612. A is correct. Bomb calorimeters measure energies; coffee cup calorimeters measure enthalpies.

613. C is correct. The pressure is kept constant by allowing gas to enter or escape the cup.

614. A is correct. Neither heat nor matter can escape.

615. D is correct. Under the constant volume conditions of a bomb calorimeter, the heat transfer corresponds to energy change rather than enthalpy change.

616. B is correct. Heat always transfers from hot to cold. The solution temperature increases as the reaction takes place and rises above the room temperature, so heat goes from the solution to the surroundings. The rise of the temperature in the solution came from bond energy.

617. B is correct. The relationship between heat capacity, change in temperature, and heat transfer is given by: $C = q/\Delta T$

618. C is correct. $C = q/\Delta T$. The combustion of 2 g of benzoic acid produces 52 kJ of energy divided by 10 °C gives 5.2 kJ/°C.

619. C is correct. $C = q/\Delta T$. 1.6 grams is 0.05 moles of hydrazine. We multiply this times -618 kJ/mol and get a heat transfer of approximately 31 kJ. The heat capacity is 6.2. Using $C = q/\Delta T$, we have ΔT equals 5. The negative sign indicates that heat evolved, so the temperature increased, as in any combustion reaction. 298 K plus 5 K is 303 K.

620. B is correct. Although water does contain covalent bonds, they don't break during melting; if they did, it wouldn't be water anymore!

621. B is correct. Ionic bonds are the only kind of bond in sodium chloride (well, there are some very weak van der Waals bonds, but they aren't offered to us as a choice).

622. A is correct. Water makes the strongest intermolecular bonds and is the heaviest. Both water and ammonia hydrogen bond, but oxygen is more electronegative than nitrogen causing the hydrogen bonding in water to be stronger. The boiling point of ammonia is -33.4 °C. Methane doesn't hydrogen bond.

623. C is correct. Dipole-dipole bonds, primarily. Carbon dioxide does contain covalent bonds, but they don't break when the CO_2 melts; if they did, it wouldn't be CO_2 anymore!

624. A is correct. For salts, the ionic bonds generally break before the covalent bonds.

625. B is correct. Diamond is a *network solid*, linked entirely by covalent bonds. Diamond cannot melt until these covalent bonds break.

626. D is correct. This is the only type of bond copper has (aside from some very weak Van der Waals bonds).

627. C is correct. Consider the intermolecular bonding in each case; the strongest intermolecular bonding should correspond to the highest boiling point. BrCl and ICl have dipole-dipole bonding, which is generally stronger than London dispersion forces (sometimes called van der Waals forces) for molecules of comparable size. So the answer should be *A* or *C*. A greater electronegativity difference increases the strength of this kind of intermolecular bond. Larger size increases the strength of London forces. Choice *C* wins on both counts.

628. B is correct. This is the only choice that explains why hydrogen fluoride is showing masses which are *multiples* of its molar mass (20 amu); dissociation would result in smaller masses. Aside from failing to explain the data, choices *C* and *D* are factually incorrect.

629. C is correct. Energy added to a solid increases vibrational kinetic energy until it reaches its melting point.

630. D is correct. Energy added to a solid at its melting point, breaks bonds and melts the solid. The vibrational energy becomes translational energy and the temperature remains constant until all the solid is melted.

631. D is correct. Energy added to a liquid at a temperature below its melting point goes into the translational and vibrational (if possible) kinetic energy of the molecules of the liquid. Breaking bonds would be associated with a phase change.

632. A is correct. Energy added to a liquid at its boiling point goes into breaking bonds. Increasing translational kinetic energy of any molecules would be accompanied by a temperature increase, which doesn't occur (ideally) until all the substance has vaporized.

633. A is correct. The heat of fusion is used when a solid changes to a liquid; vaporization is the conversion of liquid to vapor; and sublimation is the conversion of a solid to vapor. So sublimation should be the two processes together.

634. D is correct. During each process, added energy goes into breaking bonds, not increasing temperature.

635. A is correct. When heat is added to ice at 0°C, the heat will go into melting the ice, not raising its temperature. You may ask: How do we know that the ice won't all melt before we're done heating it? How do we know the temperature will still be 0°C? If you didn't have in your memory the heat of fusion of water (80 cal/g, way too big for 6 cal to melt all the ice), you'd still be able to figure it out. Here's how: if the water were already *all* liquid at 0°C, the correct answer would be *B*; therefore, since we know that we're going to use at least *some* heat to melt the ice, we know that none of the answers besides *A* will even work.

636. D is correct. One gram of H_2O would require 80 cal to melt it, 100 cal to raise the temperature to 100°C, and 540 cal to boil it. One hundred grams requires 100 times that much heat.

637. D is correct. For 1 gram, 80 calories are required to heat the water from 20°C to 100°C, but 540 calories are required to convert it to steam. Thus, more than six times the original time period is required to turn water to steam as to heat it up. Since this is true for one gram, it is true for any amount.

638. A is correct. Choices *B* and *C* are out, because evaporation is an *endothermic* process. Choice *D*, although a true statement, fails to explain the observation. We're thus left with *A*, by process of elimination.

639. A is correct. The initial dip to −35°C is probably some non-equilibrium process, since the temperature doesn't stay there for very long. The long period of constant temperature is more likely to be the freezing point of the liquid. Finally, the stabilization at −196°C is probably the temperature of the liquid nitrogen bath, since it is the lowest temperature reached.

640. A is correct. Sublimation occurs below and to the left of the triple point for any substance. Since water exists as solid, liquid, or gas at 1 atm. the triple point is below 1 atm. The triple point temperature for water is exactly 0.01 °C.

641. B is correct. The phase with the greatest specific heat requires the most heat to change its temperature by a fixed amount. This translates into the *smallest* slope on this kind of graph.

642. B is correct. The melting point is the temperature at which solid and liquid can coexist.

643. D is correct. You really need a graph with pressure, density, or volume on it to determine this.

644. A is correct. This is generally true! You can think of it according to Le Chatelier's principle, if you like.

645. A is correct. It's difficult to read precise numbers off the graph, but going from liquid to solid certainly requires a small fraction of 1000 calories; perhaps 200? But that's for 10 grams, so the heat of fusion is about 20 cal/g.

646. B is correct. Look just at the vapor part. It requires about 1000 calories to change the temperature by about 500°C. Use $Q = mc\Delta T$: 1000 = (10)c(500). Rearranging for c and solving gives us a value of 0.2.

647. D is correct. *No* substance can be brought below −273°C, since that's absolute zero.

648. A is correct. The coldest water, just turning to ice at the surface, is less dense than the warmer water, so it stays at the top of the lake. H_2O is most dense at 4°C.

649. D is correct. As you move to the left on the *x* axis, temperature is increasing. The *y* axis shows the natural log of vapor pressure. As we go up on the *y* axis, vapor pressure is increasing. To the left on the graph (at high temperatures), Z has the greatest vapor pressure.

650. D is correct. The greatest heat of vaporization is the one with the lowest vapor pressure, or, from the equation, the one with the steepest slope. The slope is equal to $-\Delta H/R$. *R* is a constant, so the magnitude of the slope is proportional to the heat of vaporization.

651. A is correct. Diethyl ether is less polar than water, so it has a lower heat of vaporization. Line W is the only one that represents a substance with a lower heat of vaporization (a less steep slope) than water.

652. A is correct. You need to memorize this.

653. A is correct. Water exists in all three phases at different temperatures at 1 atm. Only line A of the choices goes through all three phases.

654. C is correct. Only T goes through the equilibrium line between liquid and gas. This represents the boiling point.

655. B is correct. Only R goes through the equilibrium line of solid and liquid at 1 atm. Ice melts at 0 °C and 1 atm.

656. C is correct. Remember from vapor pressure that solids and liquids can be in equilibrium with their vapor pressures.

657. B is correct. Water is one of the few, but not the only, phase diagram that has a negative sloping line between solid and liquid phases on the phase diagram.

658. C is correct. The triple point is the pressure and temperature where all three phases can exist at equilibrium.

659. B is correct. When a substance changes phase from solid to gas, it sublimes.

660. A is correct. The substance will move straight up on the diagram and become a liquid from a gas.

661. D is correct. The critical temperature is the temperature above which a liquid cannot be formed regardless of the pressure applied. At temperatures above the critical temperature, gas and liquid are no longer distinct phases.

662. A is correct. When salt is added to solvent, boiling point goes up and melting point goes down. If you draw a horizontal line above the triple point to represent 1 atm, you will see that the boiling point goes up and the freezing point goes down for choice A only.

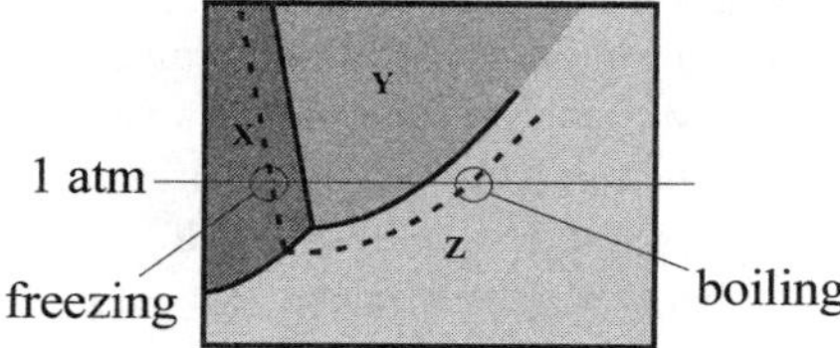

663. C is correct. This bit of trivia is slightly more than what would be required of you by MCAT, but MCAT may discuss supercritical fluid and its properties. Plasma (well beyond the MCAT) is ionized gas.

664. B is correct. As the temperature is lowered, the fluid may become a liquid.

665. A is correct. CO_2 is a gas at room temperature.

666. D is correct. Look at any phase diagram to see this.

667. D is correct. This is the definition of triple point.

668. A is correct. Choice *B* includes two sharp phase transitions, so it is out. But since beyond the critical point there is no sharp distinction between liquid and vapor, choice *A* describes a possible method for getting from liquid to vapor.

669. A is correct. The explanation says that the critical solution temperature is the temperature above which butanol is completely miscible with water for all mixtures. This is above point *A*.

670. B is correct. Point *C* is in the light region representing a one phase system. The far right of the diagram represents a mole fraction of 1 for Fluid X or pure Fluid X. Point *C* is near the right side, so the solution is mostly Fluid X at point *C*.

671. A is correct. The diagram gives the mole fraction of Fluid X as about 0.5. This just means that the entire container holds 50% Fluid X. However, from the graph, you know that it must be a two phase system. Although B says a two phase system, it goes on to describe a single phase system. From the paragraph, you know that equal amounts of Fluid X and water below the critical solution temperature form two phase systems like the one described in choice A.

672. A is correct. The experiment describes the mixture along the dotted line in the diagram.

673. B is correct. The fluid is probably somewhat polar in order to mix with water. However, since it is not completely miscible with water at room temperature, you know that it is not as polar as water.

674. B is correct. For two fluids to be completely miscible, the graph must be white for every mole fraction at a given temperature. Only graph B shows the bottom part of the graph with white space all the way across and the top part of the graph with gray space. This indicates that the fluids are completely miscible at low temperatures and only partially miscible at high temperatures. Choice C is completely miscible at high and low temperatures. Choice D is wrong because it would be impossible to have two phases if the mole fraction of one of the substances were zero.

675. C is correct. From the paragraph, the critical solution temperature is the temperature where the fluids become completely miscible. This is the highest point of the shaded region.

676. D is correct. 10° equals one. If you draw a horizontal line at 1 atm on the diagram, it passes through all phases.

677. D is correct. The triple points are where three phases exist in equilibrium. The temperatures of the three triple points are 96 °C, 119 °C, and 151 °C.

678. A is correct. Room temperature is below 40 °C. Standard state is 1 atm. Sulfur exists in its rhombic form at these conditions.

679. A is correct. The sample changes from solid to gas, which is sublimation.

680. B is correct. A phase change from gas to solid is called deposition.

681. B is correct. According to the phase diagram, sulfur melts at above 100 °C at any pressure so the water must be at a temperature greater than 100 °C. A superheated liquid is a liquid heated above is boiling point.

682. C is correct. The critical point is the greatest temperature and pressure where water and steam exist in equilibrium. This are inside the dashed line is the equilibrium area between water and steam. The highest temperature and pressure that exists in this area is given by choice C.

683. D is correct. The region enclosed by dashed line represents the equilibrium pressure and temperatures as the phase change occurs.

684. B is correct. If we follow the 300 °C isotherm, we see that a phase change occurs at 85 atm.

685. D is correct. 400 °C is above the critical temperature. By the definition of critical temperature, no amount of pressure will condense H_2O to liquid water above the critical temperature.

686. C is correct. 400 °C is above the critical temperature. By the definition of critical temperature, no amount of pressure will condense H_2O to liquid water above the critical temperature.

687. C is correct. Supercritical fluid is the fluid state at temperatures and pressures above the critical temperature and pressure.

688. D is correct. The horizontal portion is where vaporization occurs, so energy is added to break bonds. In the other portions, the temperature is constant, so the kinetic energy of the molecules is constant. *K.E.* = 3/2 *RT*.

689. D is correct. The net energy flow, ΔE, is zero because the temperature remains constant. From $\Delta E = w + q$ we see that work and heat must have opposite signs if they are not zero. There is pressure volume work done by the sample on the surroundings as the volume increases. That means that heat flow must be into the sample.

690. C is correct. Colligative properties depend upon the number of particles, not the nature or kind of particle.

691. D is correct. Colligative properties are properties that depend on the concentration of the solute particles and the nature of the solvent, but *not* on the nature of the solute particles. Heat of solution is certainly different for different solutes, and is thus not a colligative property.

692. D is correct. Remember boiling point elevation and freezing point depression.

693. D is correct. The solute molecules occupy some of the positions at the surface of the liquid and block liquid molecules from escaping the liquid. This lowers vapor pressure. Boiling occurs when vapor pressure equals atmospheric pressure, so a lower vapor pressure results in a higher boiling point.

694. A is correct. Freezing point is dependent upon symmetry as well as bond strength.

695. D is correct. Choice *A* is wrong; increasing the water content should, if anything, dilute the sugar solution and bring the freezing point back up towards zero. Choice *B* is just strange: water does not dry out its environment! Choice *C* is certainly wrong; a sugar solution will have its freezing point depressed, not raised. This leaves choice *D*. You may have been disturbed by the idea that freezing water releases heat, but this is correct; turning water into ice is an exothermic process. If you happen to live in a part of the country that suffers from freezing rain, you may note that the temperature during such events tends to climb towards 32°F. On the other hand, melting is endothermic. If snow begins to fall when the temperature is above 32°F, the snow melts upon hitting the ground, but this melting tends to lower the temperature towards 32°F, at which point the snow begins to "stick."

696. C is correct We need the highest concentration of solute particles. $CaCl_2$ breaks up into three pieces, so its particle concentration is 0.30 *M*. $NaNO_3$ breaks up into two pieces, so its particle concentration is 0.90 *M*. HCl, a strong acid, breaks up into two pieces, so its particle concentration is 1.30 *M*. Finally, NH_3 is a weak base, so it breaks up very slightly; its particle concentration is a tad more than 0.90 *M*, but certainly not as large as 1.30 *M*. Therefore, the HCl solution has the lowest freezing point.

697. A is correct. $\Delta T_f = -k_f m i$, so $([-5] - [0]) = (-1.86)(m)2$, so $m = 5/(2 \text{x} 1.86) = 1.3$.

698. D is correct. $\Delta T_f = -k_f m$, so $\Delta T_f = (-20.1)(0.01) = -0.21$. Subtracting this result from the freezing point of pure cyclohexane, which is 6.5°C, we get an answer of 6.3°C.

699. B is correct. For this problem, we use the formula $\Delta T_b = k_b m$. The *m* stands for molality of solute particles, which is defined as moles of solute per kilogram of solvent. In this case, dimethyl ether has a molar mass of 2 x 12 + 6 x 1 + 16 = 46 g, so 2.0 grams is roughly 0.04 moles. Since 100 grams of solvent weighs about 0.1 kg, the molality is 0.04/0.1 = 0.4 *m*. Plugging this result into the formula, we get $\Delta T_b = (1.71)(0.4) = 0.7$. Since this is the *change* in the boiling point, the correct answer is *B*.

700. C is correct. We know the mass of the solvent, so if we could find out the number of moles, we could find the molecular weight. Using $\Delta T_f = -k_f m$, we could find *m* from the difference in freezing point between the pure solvent and the solution if we also knew the freezing-point depression constant, k_f. Thus, *I* is certainly required, which eliminates *B* as a possibility. But recall that *m* is the molality of solute *particles*; if, for example, the solute is a strong acid, each molecule of acid would produce *two* solvated particles (one of hydronium and one of the conjugate base of the acid). Thus, we really need to know *II* as well.

701. C is correct. Plugging into $\Delta T_f = -k_f m$, we get $(-0.6) = (-5.1)(m)$, and *m* is therefore about 0.12. Since molality is defined as moles of solute particles per kilogram of solvent, and since alkanes don't dissociate, there must be about 0.12 moles of the alkane in the 1 kg of solvent. If 10 grams is equal to 0.12 moles, then the mass of one mole is about 80 grams. Since the answers are in order of increasing mass, let's start in the middle of the list of answer choices; that way we'll know whether we're too high or too low, and be able to narrow the choices down accordingly. Butane has a molar mass of 58. This answer is too low, so the right answer has to be hexane or octane. Hexane has a molar mass of 86 g/mol, which is close enough.

702. D is correct. Plugging into $\Delta T_f = -k_f m$ gives us $(-0.46) = (-1.86)(m)$, so *m* is about 0.25. Since molality is defined as moles of solute particles per kilogram of solvent, we must have 0.25 moles of solute particles in the 1 kg solvent. If choice *A* were correct, that would imply about 0.25 moles of acetic acid, since it is a weak acid and is largely undissociated. Acetic acid has a molar mass of 60 g/mol, so 0.25 moles of it would be 15 g, which is too high. Hydrochloric acid, on the other hand, would dissociate completely, and thus 0.25 moles of solute particles would come from 0.125 moles of HCl. HCl has a molar mass of 36.5 g/mol, so 0.125 moles would be about 4 or 5 grams, which is too low. NaCl, a salt, would split into two particles, so 0.25 moles of solute particles corresponds to 0.125 moles of NaCl. Since NaCl has a molar mass of 58.5 grams, 0.125 moles would be about 7 grams, which is close, but not close enough. Finally, $CaCl_2$ splits into three particles (one calcium ion and two chloride ions) and thus would need 0.08 moles to produce 0.25 moles of particles. The molar mass of $CaCl_2$ is 111 g/mol. The 0.08 moles would then be about 9 grams, and we did round some; this must be the answer. (Wow! That was a lot of work! Do MCAT questions ever require this much work? Yes, maybe once per test.)

703. B is correct. Since freezing point is higher than expected, we are looking for an answer with fewer particles. No precipitation will occur in a dilute solution of NaCl.

704. D is correct. Consider the equations $\Delta T_o = K_f mi$ and $\Delta T_c = K_f m$. If we divided the first equation by the second equation, we get: $i = \Delta T_o / \Delta T_c$.

705. B is correct. The discrepancy between the theoretical and observed van't Hoff factor is caused by ion pairing. In a more dilute solution, the ions separated by a greater distance, so there is less ion pairing.

706. C is correct. This is a somewhat unusual, but accurate, way of describing the phenomenon of osmotic pressure. Most people tend to think of the salt concentration trying to equalize, but you can just as well think of the water concentration equalizing,. In some ways, this is actually a better picture, since the water is what actually passes through the membrane.

707. A is correct. Water tends to move toward solutions with high osmotic pressure. Solute can't move through a membrane that is permeable only to water.

708. C is correct. The pressure in the pool is equal to the hydrostatic pressure only. There will be a small change in pressure because the salt solution has a greater density than the pure water, but this is not related to the osmotic pressure.

709. C is correct. The total pressure doesn't change. The osmotic pressure is the pressure necessary to bring the solution to equilibrium with pure water if the solution were separated from the pure water by a semipermeable membrane.

710. A is correct. Water flows from the inside of the cell to the outside of the cell or from low osmotic pressure to high osmotic pressure.

711. C is correct. We certainly need the osmotic pressure, which is given by MRT, or (3)(8.3)(300); don't forget to convert Celsius to Kelvin for this equation! At this point, you can already identify the answer as *C*, since *D* somehow looks like RT/M, which isn't the right expression. If you're wondering about the (1000)(9.8) that appears in every answer, it comes from physics: $P = \rho gh$, where ρ is the density of the water, which is 1000 kg/m^3.

712. D is correct. Make sure you read all the choices before selecting one! Certainly, the salt has to flow to the pure isobutanol side; there's no other possible motion! The question is: what effect will such a movement have on volume? It's possible for salt added to a solvent to actually decrease the volume, if the ions pull molecules closer together. So, although *B* is a *possible* answer, *D* is the *best* answer.

713. A is correct. There are two clues in the paragraph, first disorderly systems have lower osmotic potential, so, since a solution has greater entropy than pure water, a solution must have less osmotic potential than zero. Second, Since water moves from high to low potential, and water moves from pure water into a solution, a solution must have lower potential than pure water.

714. C is correct. The lowest osmotic potential will be the solution with the most particles. Since NaCl will dissociate, that will be the most concentrated NaCl solution.

715. D is correct. If we imagine a U tube with a selectively permeable membrane, when we apply pressure to one side, water will flow to the other side. This indicates that the side with more pressure has greater water potential.

716. B is correct. This is the definition of pH. $pH = -\log[H^+]$

717. B is correct. pH = 7 in a neutral solution only at 25 °C. B is a better definition.

718. A is correct. pH is defined by $[H^+]$. $pH = -\log[H^+]$

719. B is correct. The equilibrium constant is 10^{-14}. Hardly any of the products are formed.

720. D is correct. You should memorize A, B, and C. To help you, think about citric acid in lemons giving them a sour taste, and think about soap making bases bitter and slippery.

721. B is correct. Arrhenius acids *decrease* the pH of a solution.

722. A is correct. Arrhenius acids must produce hydronium ions *in water*. HCl certainly does, but the other two choices do not. Essentially, the laboratory definition of an acid is the Arrhenius definition. For example, if a cabinet is marked "Acids ONLY," then Arrhenius bases should not be stored in that cabinet.

723. C is correct. Brönsted-Lowry acids are capable of donating protons. Choice *III* has no protons to donate, so it cannot be a Brönsted-Lowry acid. Although ethanol is not an Arrhenius acid, since it will not donate its proton to water, it can be deprotonated by sufficiently strong bases (you see this often in organic chemistry).

724. D is correct. The Lewis definition is the most general definition of an acid. *All* Brönsted-Lowry acids are Lewis acids. In addition, anything capable of accepting an additional bond is a Lewis acid. A compound may be able to accept an additional bond because the compound has a leaving group, a double or triple bond, or, as in this case, an incomplete octet.

725. A is correct. A Brönsted-Lowry acid is a proton donor. In this reaction, NH_3 donates a proton to H^-. The fact that this reaction cannot take place in an aqueous environment (H^- is too strong a base to exist in water), and that ammonia acts as a base when it reacts with water, is irrelevant.

726. B is correct. All Brönsted-Lowry acids are also Lewis acids.

727. A is correct. The Brönsted-Lowry and Lewis definitions are in terms of a particular reaction; a substance can act as a Lewis acid in one reaction and as a Lewis base in another. Therefore, since the sign seems to be talking about substances that are always either acids or bases, it can't be using the Lewis or Brönsted-Lowry definitions. The definition the sign writer has in mind is probably the Arrhenius one.

728. C is correct. The Lewis definitions of acid and base focus on electrons. When answering questions about Lewis acids, remember that a Lewis acid is an electron-pair acceptor. An empty orbital provides an easy way to accept a pair of electrons. If you draw a Lewis structure of BF_3, you will find an incomplete octet, and therefore an empty orbital. But it's not really necessary to draw anything; process of elimination works just fine. Choice *B* would be an interesting answer, except that BF_3 is not negatively charged! (It's not written as BF_3^-.) Choice *A* is irrelevant; some fluorine-containing compounds are acids, some are bases. Choice *D* can be eliminated because BF_3 has no hydrogens, and therefore no loosely held protons.

729. D is correct. Since $pH = -\log[H^+]$, just take the power of 10 and change the sign.

730. B is correct. 3.0×10^{-4} is between 1.0×10^{-4} and $10 \times 10^{-4} = 1.0 \times 10^{-3}$, so the pH is between 3 and 4. It is not necessary to be more precise than this for the MCAT.

731. A is correct. Since the pH is between 11 and 12, $[H^+]$ is between 1.0×10^{-12} and 1.0×10^{-11}. It is not necessary to be more precise than this for the MCAT.

732. C is correct. You could probably guess this without calculation, but the calculation is also pretty easy. Every dilution by a factor of 10 increases the pH by 1 unit. Dilution by a factor of 100 increases the pH two factors of 10, and therefore the pH increases by 2.

733. A is correct. To form the conjugate base, remove a proton (remove an H^+).

734. B is correct. To form the conjugate base, remove a proton (remove an H^+).

735. A is correct. To form the conjugate base, remove a proton (remove an H^+).

736. C is correct. To form a conjugate acid, add a proton (add an H^+). Yes, this is possible in this case!

737. D is correct. Salts, strong acids, and strong bases all dissociate into ions nearly completely when dissolved in aqueous solutions; thus *B* (a salt) and *C* (a strong acid) conduct electricity quite well. Likewise, a molten (liquid) salt will conduct electricity well. But HClO is only a weak acid, and thus will not conduct electricity as well as the others.

738. D is correct. You should have memorized the fact that the first three are strong; if you *have* memorized the list of strong acids and come across an acid that is not on that list, assume it is weak, unless you are given data to the contrary.

739. D is correct. An *amphiprotic* (which means the same thing as *amphoteric*) substance is capable of gaining or losing a proton. Such is not the case for HCl, at least in water: it dissociates completely, so it loses a proton with ease, but it cannot gain a proton. Water is certainly amphiprotic, since both hydronium and hydroxide occur naturally in water. Since choice *III* must be included in an answer, the answer must be *D*.

740. D is correct. The conjugate base of the hydronium ion is water. An example where choice C is true is the conjugate base of sulfuric acid.

741. C is correct. The pH can be calculated with the hydrogen ion concentration alone: $pH = -\log[H^+]$. *A*, *B*, and *D* all require more information in order to find the hydrogen ion concentration and calculate the pH.

742. B is correct. K_a and K_b are inversely proportional, but, because K_w is small, it is possible for K_a and K_b to be small and for a weak acid to have a weak conjugate base. However, a strong acid must have a weak conjugate base and a strong base must have a weak conjugate acid.

743. A is correct. This can be seen most clearly by looking at the equation: $HA + H_2O \rightarrow A^- + H_3O^+$. In the forward reaction HA transfers a proton to H_2O; in the reverse reaction, H_3O^+ transfers a proton to A^-. If HA is a strong acid, the equilibrium favors the forward reaction.

744. B is correct. The definition of a strong acid is an acid that is stronger than the hydronium ion. You should recognize the other choices as strong acids.

745. D is correct. The acidity of the hydrogen halides increases moving down the periodic table.

746. D is correct. A, B, and C are very strong bases.

747. C is correct. $Mg(OH)_2$ is insoluble and thus not a strong base. The others are strong bases.

748. A is correct. It is easiest to compare the conjugate acids here: HClO, $H3O^+$, and HCl. HCl is a strong acid and HClO is a weak acid. By definition, a strong acid is a stronger acid than $H3O^+$ and a weak acid is weaker than $H3O^+$. The strength of their conjugate bases will be in reverse order.

749. D is correct. The percent ionization is not the same as acid strength. It depends on the factors listed.

750. D is correct. Stronger acids dissociate to a greater extent, but even strong acids can't dissociate completely in very concentrated solutions. Increasing temperature usually increases the percent dissociation .

751. C is correct. Proton transfer indicates an acid-base reaction. A pH change of one point is equivalent to a proton concentration change by a factor of 10. From the rate law for an acid-base reaction (rate = $k[H^+][\text{base}]$), we see that the rate is proportional to the proton concentration.

752. B is correct. The definition of pH is the negative log of the H ion concentration.

753. B is correct. One pH point is equal to a tenfold change in the hydrogen ion concentration.

754. D is correct. With each pH point, the hydrogen ion concentration increases by a factor of ten.

755. C is correct. Answers A, B, and D are false.

756. D is correct. The strength of an oxyacid increases as the number of oxygens surrounding it increases. Another way of stating this is answer choice D.

757. A is correct. Choice A is an ionic oxide. Ionic oxides form bases.

758. D is correct. The strength of an oxyacid increases as the number of oxygens surrounding it increases.

759. A is correct. On an oxyacid, when the central atom is different and each central atom has the same number of oxygens, the central atom with the greatest Electronegativity produces the strongest oxyacid.

760. D is correct. The oxygens are electron withdrawing, which tends to polarize the bond on the acidic hydrogen increasing its likelihood of dissociating in water.

761. B is correct. There are three major factors affecting acid strength, H–X bond strength and polarity and the stability of the conjugate base. The weaker the bond, the stronger the acid, and the more polar the bond the stronger the acid. A stable conjugate base creates a stronger acid.

762. C is correct. A hydride contains hydrogen and one other element.

763. B is correct. None of these hydrides can hydrogen bond, so a general boiling point comparison can be made based upon molecular weight. The greater the molecular weight, the greater the boiling point.

764. C is correct. Think about HCl, HBr and H_2S. The hydrogen halides are stronger acids than H_2S, and their strength increases as you go down the periodic table. You could also compare water to H_2S. H_2S is a stronger acid than water.

765. D is correct. Metal hydrides like NaH are basic or neutral, while nonmetal hydrides like H_2O and HCl are acidic or neutral. Ammonia, NH_3, is an important exception to this rule.

766. D is correct. Acid strength of the hydrides increases as you go down the periodic table.

767. C is correct. The ionic hydrides are bases.

768. D is correct. The hydride ion is a stronger base than the hydroxide ion. HClO is a weak acid.

769. D is correct. The hydride ion is a stronger base than the hydroxide ion. This pushes the equilibrium to the right.

770. B is correct. Students are often surprised by this one. Consider the mathematics of it: if HF has a pK_a of 3.1, its conjugate then has a pK_b of 11. Even though the acid is weak, the conjugate base is also weak.

771. B is correct. The K_a is just K_w divided by 55 the concentration of water (55 mol/L).

772. A is correct. Don't do any math here. The equilibrium for the autoionization of water lies far to the left.

773. D is correct. Answers A, B, and C are related to thermodynamic properties. The value of a thermodynamic property does not affect rate.

774. C is correct. The strongest base has the smallest pK_b (remember that there is a minus sign in the definition of pK_b). Using the fact that pK_a + pK_b = 14 for a pair of conjugates, we can see that SCN^- has a pK_b of 15.8, BF_4^- has a pK_b of 13.5, and IO^- has a pK_b of 3.5. So out of this group IO^- is the strongest base. What about Cl^-? Although we don't know the exact pK_a of HCl, its conjugate acid, we do know HCl is a strong acid. Strong acids generally have negative pK_a's, so its conjugate should have a pK_b above 14, which would make it weaker than IO^-.

775. D is correct. The pK_a is the *negative* log of K_a, so choices *II* and *III* correspond to acids with equilibrium constants less than one. If the equilibrium constant is less than one, the acid is less than half dissociated at equilibrium.

776. C is correct. A tricky one. For an acid HA, $K_a = -[H_3O^+][A^-]/[HA]$. For water, since $[OH^-] = 1 \times 10^{-7}$, we know that $K_a = (1 \times 10^{-7})(1 \times 10^{-7})/55.6$, or between 10^{-15} and 10^{-16}. Since p$K_a = -\log K_a$, we get a pK_a of between 15 and 16. Note that the K_a of water is somewhat differently defined than K_w, which is just $[H_3O^+][OH^-]$, without the $[H_2O]$ in the denominator.

777. C is correct. For conjugates, pK_a + pK_b = 14.

778. B is correct. For conjugates, pK_a + pK_b = 14.

779. D is correct. The equation pK_a + pK_b = 14 applies to *conjugates* only. The pK_b of the carbonate ion, which is conjugate to the hydrogen carbonate ion, is indeed 3.75, but that is not what this question asks for!

780. D is correct. This answer may surprise you, but trust the math. If benzoic acid has a pK_a of 4.19, then its conjugate has a pK_b of 9.81, making it a weak base.

781. C is correct. Many students are afraid of these "Roman numeral" problems, but they are really easier than a normal question because they allow you to use process of elimination. For example, you might know that methods *III* and *IV* will certainly work. After all, method *III* gives the pK_a directly; using method *IV*, all a scientist would have to do is subtract from 14. This already limits our choices to *C* or *D*. Therefore, we don't have to think further about *II*; apparently, it is a valid method! But we do have to consider *I*. The amount of base required to neutralize the acid is given by $M_1V_1 = M_2V_2$. None of these factors relate to the pK_a, so method *I* shouldn't work. (In fact, method *I* cannot even distinguish between weak and strong bases.) Thus, we're only left with choice *C*.

782. C is correct. Equilibrium in an acid/base reaction will always lie on the side of the reaction where we find the weaker acid (which is the same as the side with the weaker base). Since higher pK_a's represent weaker acids, HCN is the weaker acid, and the equilibrium will thus lie further toward the left side of the equation.

783. C is correct. HBr is a strong acid, so it dissociates almost completely, giving an H_3O^+ concentration of 0.01 *M*. Since pH = $-\log[H_3O^+]$, the pH is 2.

784. D is correct. KOH is a salt, and thus dissociates completely, giving an OH^- concentration of 0.01 *M* and a pOH of 2. But, we know that pH + pOH = 14, so the pH = 12. Or, skip the calculations and realize that choice *D* is the only basic pH!

785. C is correct. See the previous problem.

786. B is correct Almost no calculation is necessary. First of all, since acetic acid is an acid, the resulting pH must be less than seven, eliminating choices *C* and *D*. If acetic acid were a strong acid, a 0.1 *M* solution would have a pH of –log 0.1, or 1.0, which is choice *A*. But acetic acid is a weak acid, so the pH must be not quite so acidic. The answer is *B*.

787. A is correct. HCN is an acid, so the pH must be below 7. Since there is only one answer below 7, you do not need to do any calculation. This sort of "shortcut" is encouraged by the people who write the MCAT, so look for it!

788. A is correct. HCl is a strong acid, so it dissociates completely. The H_3O^+ concentration is thus 7 *M*, and the pH = –log 7. Since log 1 = 0 and log 10 = 1, the pH is between 0 and –1.

789. D is correct. If the acid created a good buffer at 6.5, there will be virtually no change in pH until the buffer is extremely diluted. If the solution is not buffered, the acid will dilute and move toward a pH of 7. Adding water to an acid could never raise the pH above 7.

790. B is correct. An aqueous HCl solution cannot be basic; however, this solution is so dilute that we must consider the hydrogen ions donated by water. The total number of hydrogen ions in solution are those donated by water, 10^{-7}, and those donated by HCl, 10^{-8}. The sum of these is $1.1\text{x}10^{-7}$. The negative log of $1.1\text{x}10^{-7}$ is 6.96.

791. D is correct. You cannot know the pH of these solutions without knowing the concentrations of the acids.

792. D is correct. If you know *I* and *II*, you can calculate the pH of the solution.

793. C is correct. *Don't think about this until you look at your choices!* You may come up with a perfectly reasonable explanation, but it may not be the one the question-writer had in mind. Instead, use process of elimination. You can eliminate choices either because they are factually incorrect or because they have nothing to do with the phenomenon described. Taking the choices one at a time: choice *A* would be an explanation if it were true, although a hydrophobic residue would not activate an acid. Choice *B* is factually untrue (many proteins are found in nearly neutral environments), and it also wouldn't explain the phenomenon (an acidic environment makes weak acids *less* likely to dissociate, by Le Chatelier's principle). Choice *C* would be an explanation if it were true; plus, since basic amino acids are stronger bases than water, this looks like a good answer. In considering choice *D*, realize that at high concentrations acids tend to be less dissociated, at least on a percentage basis. At any rate, weak acids do not dissociate to a very great extent, even at low concentrations.

794. D is correct. Although we know the pK_a of this substance, we do not know its pK_b. We cannot simply subtract its pK_a from 14, since doing so would give the pK_b of the substance's conjugate. Therefore, since we don't know whether this substance is a more effective acid or base, we cannot determine what the pH will be.

795. D is correct. 5.0 L is one hundred times bigger than 50 mL, so the molarity of the nitric acid should drop by a factor of 100. Since nitric acid is a strong acid, it completely dissociates, and the hydronium concentration drops by a factor of 100 as well. According to the definition of pH, each drop by a factor of 10 in hydronium concentration causes the pH to go up by one. Since 100 is equal to 10 to the second power, the pH rises from 1.3 to 3.3.

796. D is correct. As usual with "explanation" questions, use process of elimination to get rid of answers that are either factually incorrect or don't explain the phenomenon. Choice *A* is factually incorrect. So is choice *B*; they are strong acids! Choice *C* sounds plausible for strong acids, but so does choice *D*. Which is really the important factor? Consider a typical weak acid, like acetic acid. It is stronger than water, but does not dissociate completely. Thus, the key cannot be strength relative to water, leaving *D* as the answer.

797. C is correct. In aqueous solution, water eagerly accepts a proton from either HCl or HI, and we get total dissociation. Since HI produces more H^+ in acetic acid solution, HCl must not totally dissociate in acetic acid. This can only mean that acetic acid is less willing to accept protons than water. Accepting protons is a basic quality. Water is a stronger base than acetic acid. Choice A is true, but doesn't explain anything. Choice B is false.

798. B is correct. Choice *A* is factually incorrect: alcohols are more water-soluble than alkanes. Choice *B* seems plausible. Choices *C* and *D* are factually incorrect; both are effectively neutral in aqueous solution.

799. B is correct. Since hydrochloric acid is a strong acid, it dissociates completely, and the resulting chloride ion must have no tendency to pick up a proton in aqueous solution. Thus, sodium chloride should be neutral.

800. A is correct. The conjugate of a weak base will be more acidic than the conjugate of a strong acid is basic. The solution will have a pH below 7. Imagine the conjugate of HCl and the conjugate of NH_3. The salt formed is NH_4Cl.

801. D is correct. All cations except those of alkali metals and the heavy alkaline earth metals (Ca^{2+} and heavier) form weakly acidic solutions in water. Since all the choices were alkaline earth metal cations, you should have at least narrowed the answer choices to C or D, the heaviest and lightest of the alkaline earth metal family choices given.

802. B is correct. All cations except those of alkali metals and the heavy alkaline earth metals (Ca^{2+} and heavier) form weakly acidic solutions in water.

803. C is correct. Cations that form acidic solutions do so upon hydration. Their positive charge withdraws electrons from the oxygen in water weakening the O–H bond in water. The greater the charge of the cation and the smaller the size, the stronger the acidic solution formed. Na^+ does not form an acidic solution in water. All cations except those of alkali metals and the heavy alkaline earth metals (Ca^{2+} and heavier) form weakly acidic solutions in water.

804. B is correct. This is the hydrolysis of iron hydrate, not water.

805. A is correct. The greater the charge of the cation and the smaller the size, the stronger the acidic solution formed. Na^+ does not form an acidic solution in water. Of the choices given, Fe^{3+} is the smallest ion with the greatest charge.

806. B is correct. Since acetic acid is a weak acid, it dissociates to only a small amount, indicating that the resulting acetate ion can easily pick up a proton again. This suggests that the acetate ion is basic, and thus so is sodium acetate.

807. C is correct. Nitrate is conjugate to nitric acid, a strong acid, and is thus neutral. Cyanide is conjugate to hydrocyanic acid, a weak acid, and is thus a weak base. Carbonate is conjugate to the hydrogen carbonate ion, which in itself is conjugate to carbonic acid. Carbonic acid is a weak acid, therefore its conjugates are bases. Remember, memorize the common strong acids, and assume other acids are weak unless told otherwise.

808. C is correct. This salt is made from the conjugates of a weak acid and a weak base. Whether the pH is acidic or basic depends upon the strengths of the conjugates. The K_a is an indicator of the acid strength and the K_b is an indicator of the base strength. The K_b is larger, so the pH will be over 7.

809. D is correct. If the salt is more soluble in basic solutions, it must be acidic. (Bases react with acids, so an acidic salt would constantly find its ions in solution disappearing. Le Châtelier's principle would then drive the solubility reaction forward.) NH_4^+ is acidic, since it is conjugate to the weak base ammonia. Chloride, of course, is neutral, being conjugate to the strong acid HCl. Thus, the answer is *D*.

810. D is correct. As H_3PO_4 loses protons, it becomes a weaker and weaker acid.

811. C is correct. HCO_3^- is amphoteric, so we have to consider its ability to act as both acid and base. As acid, the pK_a is 10.25, since that would be the second dissociation of carbonic acid. As a base, it is conjugate to carbonic acid, and would thus have a pK_b of 14-6.37, or about 8. Thus it is a "better" base than acid, and the pH would be greater than 7.

812. C is correct. Titrations allow you to find the concentration of an unknown solution.

813. B is correct. The equivalence point is 50 mL for all six titrations. At the equivalence point the number of base molecules equals the number of acid molecules. $50\ mL_{NaOH} \times 0.1\ mol/L = 50\ mL_{acid} \times C_{acid}$.

814. C is correct. To find the K_a, look at the half equivalence point, which occurs halfway to the equivalence point or at 25 mL in this case. At the half equivalence point, the pH equals the pK_a, so the pK_a is 4 for acid T. $pK_a = -\log(K_a)$. If the pK_a is 4, the K_a is 1×10^{-4}.

815. B is correct. The pH is 3. $pH = -\log[H^+]$.

816. C is correct. The pH of the conjugate of S can be found by looking at the equivalence point. At the equivalence, only the conjugate base of S is left. However, this solution has been diluted by 2 by the titrant, so the equivalence point reflects the pH of a 0.05 *M* solution of the conjugate base. The equivalence point is near 9 on the graph. You could also calculate exactly what the pH of 0.05 M conjugate of S would be by using the pK_a of S. From the pK_a find the pK_b and calculate from there. From the half equivalence point, we know that the pK_a is 6, so the pK_b is 8. K_b is 10^{-8}. $K_b = [OH^-][HA]/[A^-] = x^2/0.05$

817. C is correct. The endpoint range should cover the equivalence point, which occurs here at 50 mL.

818. A is correct. The pH is about 4, so the concentration of H^+ is about 10^{-4}. This is the concentration of dissociated acid. The concentration of undissociated acid is the original concentration minus this: 0.1 – 0.0001. This is approximately 0.1. 0.0001/0.1 = 0.1 %.

819. B is correct. The vertical line on the graph goes through all these pH points. However, from the starting pH of 1 you know that acid V totally dissociates; it is a strong acid. The titrant is a strong base, so the pH of the equivalence point must be 7.

820. C is correct. This is just curve S going from right to left. Acid V is a strong acid. The conjugate of S is a weak base. When a base is titrated with acid, the half equivalence point is when the concentration of the conjugate acid equals the concentration of the base: $[HA] = [A^-]$. When an acid is titrated by a base, the half equivalence point is when the concentration of conjugate base equals the concentration of acid: $[A^-] = [HA]$. These are the same positions. The half equivalence point on S is below 7, so C is correct.

821. B is correct. The starting pH is below 7, so this is a weak acid. It is clearly not water. You can check this by choosing any point on the graph and calculating the pH for the corresponding concentration of NaOH.

822. C is correct. Look at the half equivalence points. An acid buffers best at its half equivalence point.

823. A is correct. Using $M_1V_1 = M_2V_2$, we substitute:

$$(5.0)(25) = M_2(50)$$

Note that there is no need to convert out of milliliters in this formula, since the conversion factor would affect both sides of the equation equally.

824. A is correct. Using $M_1V_1 = M_2V_2$, we substitute:

$$(7.0)(30) = M_2(100)$$

Note that there is no need to convert out of milliliters in this formula, since the conversion factor would affect both sides of the equation equally.

825. D is correct. All titration curves are roughly S-shaped, eliminating choices *A* and *B*. Since we are starting with a base and adding acid, the pH should drop. That leaves only choice *D*.

826. D is correct. Since sodium hydroxide is a strong base, you do not initially need to consider equilibria, but you can instead do this as a limiting reagent problem. 30 mL of 3 *M* acetic acid is 30 x 3 = 90 millimol. 50 mL of 2 M sodium hydroxide is 50 x 2 = 100 millimol of base. So there is more base than acid, and some of the base will be left over. Therefore the final pH will be greater than 7.

827. D is correct. Since sodium hydroxide is a strong base, you do not initially need to consider equilibria, but you can instead do this as a limiting reagent problem. 30 mL of 3 *M* acetic acid is 30 x 3 = 90 millimol. 45 mL of 2 *M* sodium hydroxide is 45 x 2 = 90 millimol of base. Since there are equal amounts of base and acid, both reactants will be gone. The products of this reaction are water and (sodium) acetate. Since acetate is the conjugate of a weak acid, it is a weak base. Thus, the solution will be basic and the pH will be greater than 7.

828. B is correct. We start the same way we started questions 496 and 497: 30 mL of 3 *M* acetic acid is 30 x 3 = 90 millimol. 40 mL of 2 *M* sodium hydroxide is 40 x 2 = 80 millimol of base. Since there is more acid than base, the acid will be left over. This time it is only a weak acid, however, and one of the products will be its conjugate base. Thus, a buffer has been formed. The pH of a buffer is generally fairly close to its pK_a (it might be off by one or two pH points), so the pH should be somewhere between 3 and 6.

829. A is correct. Using $M_1V_1 = M_2V_2$, we substitute:

$$M_1(30) = (7.0)(21)$$

Note that there is no need to convert out of milliliters in this formula, since the conversion factor would affect both sides of the equation equally.

830. D is correct. First, note that all the answers include water and sodium ions, so we need not consider these species when trying to determine the answer. The reaction taking place is:

$$OH^- + HAc \rightarrow H_2O + Ac^-$$

where Ac^- represents the acetate ion. So clearly some acetate ion is present, and the answer must be *A* or *D*. Since acetic acid and hydroxide react with each other, answer *A* doesn't make sense, leaving *D* as the correct answer. If you wanted to, you could also answer this question by noting that there are fewer moles of sodium hydroxide than of acetic acid. Thus, hydroxide is the limiting reagent in the above reaction, and would not be present in the final solution. Acetic acid would be in excess, and would thus be present. This again points to *D* as the correct answer.

831. A is correct. An aqueous buffer is composed of a weak acid and its conjugate base (or a weak base and its conjugate acid, which is really the same thing). HCl is a *strong* acid and therefore cannot be used in a buffer in aqueous solution.

832. B is correct. Here's the reaction: $HCl + NH_3 \rightarrow Cl^- + NH_4^+$. It is correct to say that the reaction goes nearly to completion because HCl is strong; it would only be represented as an equilibrium if *both* were weak. Since ammonium is the conjugate of a weak base, it is a weak acid. Chloride, on the other hand, is neutral (it can't pick up a proton, since HCl dissociates completely in water). Thus, the resulting solution is acidic.

833. C is correct. Ammonium is the conjugate acid of ammonia, which is a weak base. Therefore, this is a buffer. For buffers *only*, you can calculate the pH by using the following equation: pH = pK_a + log ([base]/[acid]). In this case, since [base]=[acid] and log 1 = 0, pH = pK_a.

834. D is correct. Again, this is a weak base and its conjugate acid, and is therefore a buffer. For buffers *only*, you can calculate the pH by using the following equation: pH = pK_a + log ([base]/[acid]) = 9.26 + log (2/.2) = 9.26 + log(10) = 9.26 + 1 = 10.26.

835. A is correct. Careful; HCl is a strong acid, so this is *not* a buffer; however, the sodium chloride solution does dilute the HCl by a factor of 2. The HCl completely dissociates, giving $[H_3O^+] = 1$ *M*. Diluted by the sodium chloride solution makes it a 0.5 molar solution. Then, pH = $-\log[H_3O^+] = -\log(5\text{x}10^{-1}) = 0.3$. This answer is still closest to zero.

836. D is correct. This problem looks hard at first, but it's really not so bad. The usual form of the Henderson-Hasselbalch equation involves pK_a, but we want to use the pK_b of the conjugate base. Since we know that pK_a + pK_b = 14 for conjugates, then pK_a = 14 – pK_b. We can substitute this form into Henderson-Hasselbalch to get choice *D*. Also, you may have noticed that choices *B* and *C* are algebraically identical; therefore, neither can be correct.

837. B is correct. Recall from the Henderson-Hasselbalch equation that the pH of a buffer depends on the ratio of base to acid. Diluting the solution does not change the pH.

838. C is correct. The pH of a buffer depends only on the pK_a of the acid and the ratio of base to acid. Since the ratio is the same in both cases, the pH is the same. But solution A has a greater capacity to absorb any acid or base that is added to the solution.

839. B is correct. The equivalence point can be found from the titration curve. This gives the volume of base required to titrate the acid, and, since the concentration of base is known, the original concentration of acid. Since the acid was weak, there will be a buffer region before the equivalence point, and, in the middle of the buffer region, pH = pK_a. From the titration curve alone, however, there is no way to determine the acid's molecular weight.

840. C is correct. At the equivalence point, the acid has been converted entirely into its conjugate base, which will be weakly basic.

841. D is correct. At this point, pH = pK_a, but for some weak acids the pK_a is below 7, for others it is above.

842. B is correct. The indicator should change color near the equivalence point.

843. D is correct. The titration is considered complete when the indicator changes color. This is called the endpoint. The equivalence point is defined as the point when the amounts of reactants are equal. When a titration is performed, the experimenter generally *assumes* that the equivalence point is near the endpoint, but it's not guaranteed.

844. C is correct. Bromthymol blue has a pK_a of 7.1. This is the closest to the equivalence point.

845. C is correct. Think about $M_1V_1 = M_2V_2$ if you like; there's no room for acid strength to play a role in the volume calculations.

846. C is correct. It is best to count equivalence points when confronted with a polyprotic acid. There are only two equivalence points, so the correct answer is *C*.

847. B is correct. Since the titrant and the sample have the same molarity, the volume of the sample is equal to the volume of the titrant at the first equivalence point.

848. A is correct. A more concentrated acid would have a lower pH. The equivalence point is found from the concentration of conjugate base, so the conjugate base would be more concentrated at the equivalence point.

849. D is correct. You should recognize the second equivalence point as the second vertical region on the curve.

850. C is correct. This is the second half equivalence point which is the pK_{a2}. You can calculate it or read it from the curve.

851. B is correct. Adding a small amount of water to a buffered solution will not significantly change the pH.

852. B is correct. The equivalence point is 9.5. This corresponds to $10^{-9.5}$ *M* hydrogen ion concentration.

853. A is correct. There is virtually no carbonic acid or bicarbonate ion because it has all been neutralized by NaOH. The carbonate ion is not at 0.1 molar concentration, because the volume has been increased by a factor of 3 by the NaOH solution.

854. C is correct. Using $M_1V_1 = M_2V_2$ gives the first equivalence point. In this case, $(3.00)(50.0) = (7.00)V_2$, so V_2 is a bit more than 20 mL. But at the first equivalence point, only one proton per molecule has been removed, and two are required to completely convert carbonic acid to carbonate. Therefore the answer is twice this amount.

855. D is correct. At the first equivalence point, the dominant species will be amphoteric. Some amphoteric species are Arrhenius acids (HSO_4^-, for example), while some are basic (like HCO_3^-).

856. C is correct. At the second equivalence point, the diprotic acid will have been converted into the entirely deprotonated form, which is certainly basic.

857. D is correct. A and D contradict each other. K_a values for successive protons on a single polyprotic acid decrease.

858. D is correct. We use the equilibrium expression and set the K_a equal to $x^2/1\text{-}x$. We assume that x is small compared to 1 and throw it out of the denominator. x is the concentration of hydrogen ions. x equals about $1.1\text{x}10^{-1}$, which is much smaller than the original 1 mole of HSO_4^-, so some, but not most, of the ions lose a proton. (By the way, the value of x was over 5% of 1, so throwing x out of the denominator made our calculations pretty rough. Typically, you can only throw out the x if it is below 5%.)

859. B is correct. This is the half equivalence point. At the half equivalence point, the acid and its conjugate base have equal concentrations.

860. A is correct. The equivalence points depend upon concentration.

861. A is correct. In a redox reaction, a compound is the reducing or oxidizing agent, and an atom in those compounds is oxidized or reduced.

862. B is correct. The compound is the oxidant or reductant, and the atom is reduced or oxidized.

863. D is correct. Lose electrons = oxidation; gain electrons = reduction.

864. D is correct. Redox reactions do not necessarily represent actual transfers of electrons. For instance, the charge on an atom does not necessarily change. Instead, oxidation states represent a model to help understanding of redox reactions.

865. D is correct. The oxidizing agent takes electrons from the atom that is oxidized, so an atom in the oxidizing agent is reduced.

866. A is correct. When doing this type of question, it's helpful to check for the quicker possibilities first. The reaction is clearly not a combustion (a combustion is simply a combination with oxygen). It is also not Brönsted-Lowry acid/base, since what is transferred is an H^-, not an H^+. We're left with redox or Lewis acid/base. Of the two, redox is easier to check; Lewis acid/base requires sketching several Lewis structures. The oxidation numbers are given below:

$$Ca^{(+2)}\,H^{(-1)}{}_2 + 2B^{(+3)}H^{(-1)}{}_3 \rightarrow 2B^{(+3)}H^{(-1)}{}_4{}^- + Ca^{2+}$$

As can be seen, the oxidation numbers do not change. Therefore, by process of elimination, we learn that the reaction is Lewis acid/base.

867. A is correct. Protons are transferred from the sulfuric acid to the fluoride ions. However, you should always check the other answers if you have time. It's not redox, because you can see the acidic hydrogens remaining acidic, the calcium ion remaining a calcium ion (it must be +2), and the sulfate and fluoride ions retaining their identities. It's not combustion, because combustion requires oxygen as a reactant. And it's not decomposition, because there's more than one reactant.

868. C is correct. In all of the steps, the oxidation number of at least one of the elements changes.

869. D is correct. Since each oxygen is –2, the oxygens add up to –6. The hydrogen is +1, and we know the entire molecule must add up to zero, since it has zero charge.

870. D is correct. This is a salt (metal and nonmetal), so we should use the charges on the ions. Sulfur is –2 (same column as oxygen), so the iron must be +3.

871. B is correct. Phosphate is PO_4^{3-}. Each oxygen is –2, for a total of –8. Since the total must be the charge on the ion (–3), the phosphorus must be +5.

872. C is correct. Sodium dichromate is a salt, so first break it into ions. The sodium ion is +1, so dichromate must be $Cr_2O_7^{2-}$. Each oxygen is –2, for a total of –14. Thus the chromiums must add up to +12, and each chromium is +6.

873. B is correct. This compound is ionic, so you can break it into K^+ ions and what is evidently a XeO_6^{4-} ion. Since each oxygen has an oxidation number of –2, and the xenon and six oxygens together must total to –4, the xenon must be +8. Since xenon is a noble gas, it has 8 valence electrons, and a +8 oxidation number is acceptable.

874. B is correct. For pure elements, the oxidation number is always 0.

875. C is correct. Sodium azide is a salt, so you should first break it into ions. The sodium ion is +1, so the azide ion must be –1. But this ion contains three nitrogens, so each nitrogen is –1/3.

876. C is correct. How does this relate to redox? Oxygen's strong electronegativity helps to explain why O_2 is a strong oxidizing agent.

877. D is correct. Since fluorine is more electronegative, we have the very unusual circumstance where oxygen is in an oxidized state. If this question were an MCAT question, you would have to ask yourself which rule should you follow and which should you break. 1. Oxygen has an oxidation state of –2 except when it is in a peroxide or in its standard state. or 2. Fluorine is the most electronegative element. The first rule is a memorization technique, while the second is a basic scientific fact. The second rule wins. Memorization techniques can get you into trouble if you rely on them exclusively.

878. C is correct. Like the previous question, we have the very unusual circumstance where oxygen is in an oxidized state.

879. C is correct. The oxidation numbers must add up to the charge on the molecule. Each oxygen will have the same number, so each oxygen must be –1/2.

880. B is correct. In reaction B Oxygen is oxidize and reduced. It begins with an oxidation number of –1 in hydrogen peroxide and ends with an oxidation number of zero in oxygen gas and –2 in water.

881. A is correct. A basic solution would remove H^+ ions and pull both half reactions to the left. Hydrogen peroxide is a reactant in R1 and donates electrons, so it is the reducing agent. In R2, hydrogen peroxide is a product going to the left, and is neither the reducing nor the oxidizing agent in that case. R2 is pushed to the right by acid, indicating that in acidic solutions hydrogen peroxide is an oxidizing agent.

882. C is correct. This compound is a salt, so first break it into ions: K^+ and HSO_4^-. Each oxygen is –2 for a total of –8, and the H is +1, so the sulfur must be +6.

883. A is correct. Of the choices, chromium has the lowest oxidation state in CrO.

884. D is correct. Oxygen has an oxidation number of –2 in all these cases. Don't forget that oxidation number is determined *per atom*, so, for example, nitrogen in N_2O has an oxidation number of +1.

885. C is correct. Since we know that oxygen prefers to have an oxidation number of –2, it must somehow be forced into other values. One way to do this would be to have it be the anion in an ionic compound, since the sum of the oxidation numbers *must* be equal to the charge on the ion. Choices *B* and *C* look promising in this respect. An examination of choice *B*, however, reveals that oxygen has an oxidation number of –2 for the first three compounds. But in choice *C*, the oxidation number drops from –2 to –1 to –1/2 (remember, oxidation number is calculated per atom) to 0, as required.

886. B is correct. First of all, "agents" are always reactants, so the correct answer must be *A* or *B*. The reducing agent is the compound that contains the element being oxidized, which in this case is iron.

887. B is correct. The reducing agent is the compound that contains the element that is oxidized. Iodine in FeI_2 has an oxidation number of –2, but I in I_2 must have an oxidation number of 0. Therefore, since the oxidation number of iodine goes up, it is oxidized, and FeI_2 is the reducing agent.

888. D is correct. The reducing agent is the compound that contains the element that is oxidized. The halogens on each side remain at –1, so the Fe and V must be +2 in all cases.

889. B is correct. Only the oxidation number of the sulfur in SO_2 increases, from +4 on the left to +6 on the right.

890. A is correct. The chlorine is reduced from +5 on the left to +4 on the right.

891. B is correct. The reducing agent is the compound that contains the element that is oxidized. Manganese is +2 on the left and +3 on the right, so it is oxidized; therefore, MnS is the reducing agent.

892. D is correct. Water can also act as a Brönsted-Lowry acid, a Lewis base, and a reducing agent. Quite a versatile substance!

893. B is correct. No oxidation numbers change, which eliminates choices *C* and *D*. The iron has empty orbitals that the lone pairs on the oxygen attach to, which makes it a Lewis base.

894. C is correct. The hydrogen in water is reduced from +1 to 0, so water is an oxidizing agent. You might be tempted to think that this is a Brönsted-Lowry reaction, since one of the water molecules became a hydroxide ion, but it's not: the positive charge ended up on the sodium, not the hydrogen, so this isn't an acid-base reaction.

895. A is correct. Hydrogen in water has an oxidation number of +1, while in hydrogen it is 0. Thus, the hydrogen in the water is reduced, and the water is an oxidizing agent.

896. D is correct. The oxidation numbers of neither hydrogen nor oxygen change in this reaction.

897. C is correct. Nitrogen has an oxidation number of +4 in NO_2. That increases to +5 in HNO_3 but decreases to +2 in NO. This kind of reaction, where the same compound is both the oxidizing and reducing agent, is called a *disproportionation* reaction.

898. D is correct. The nitrogen is reduced from +5 in HNO_3 to +3 in HNO_2.

899. A is correct. There is no change in oxidation number; the ion just needs to be protonated.

900. D is correct. The nitrogen is reduced from 0 to –3. (If you marked *A*, you're confusing neutral hydrogen with protons.)

901. D is correct. Although oxidation numbers are not necessarily charges, they can never be greater than the number of valence electrons the element possesses (note: the *d* and *f* electrons of transition metals are not technically valence electrons, but should be treated as such for this rule). Since carbon has 4 valence electrons, its maximum oxidation number is +4.

902. C is correct. Voltage is relative. It is a comparison of electrical potential between two objects or solutions. Like gravitational potential, it must be measured from an arbitrary point, such as height from sea level, or from the ground, etc. In a redox titration, the voltage is measured from a solution called a standard.

903. C is correct. Clearly the answer must be A, B, or C because they cover all the possibilities. A redox reaction can be a spontaneous or nonspontaneous process resulting in higher or lower electrical potential; it depends upon the reaction.

904. A is correct. The paragraph states that potassium permanganate is purple. When the permanganate ion is dripped into the Fe^{2+} solution, MnO_4^- is immediately reduced to Mn^{2+}. Mn^{2+} is colorless. Once all the iron ions are used up, no more oxidation occurs, and the solution turns purple with the MnO_4^- ions.

905. C is correct. The paragraph states that Fe^{2+} is already in the solution, so it must be a reactant. Since MnO_4^- is reduced to Mn^{2+}, Fe^{2+} must be oxidized. Iron will not take on a plus 4 oxidation state.

906. A is correct. The permanganate ion is reduced. Choice C is wrong because it is oxidation. Choices B and D are wrong because there is no Mn^{7+} or Mn^{7+} in solution.

907. C is correct. Choice C is the only reaction balanced in matter and charge. It is also the only reaction that takes place in acidic solution.

908. D is correct. The half reaction for iron involves the transfer of one electron and the half reaction for the permanganate ion involves the transfer of five electrons. We use the lowest common multiple, which is five.

909. D is correct. Since Fe^{2+} loses one electron and the permanganate ion gains 5 electrons, five Fe^{2+} ions are required for the reduction of each permanganate ion.

910. B is correct. The number of moles of permanganate ion in solution is 0.02 L x 0.01 mol/L = 0.0002 mole. Five iron ions were required to reduce each permanganate ion, so the number of iron atoms in solution was five times the number of permanganate ions or 0.001 mole.

911. B is correct. The previous question tells us that there were 0.001 moles in solution. 0.001 moles/0.04 L = 0.025 *M*.

912. B is correct. This question requires too much calculation to be on the MCAT. But the previous questions lead up to it and you should be able to do it anyway. There are 0.001 mole of iron times 55.85 g/mol. This gives about 0.056 grams of iron. The mass of iron divided by the mass of the iron ore gives the percent mass. 0.056g/0.56g = 0.1 = 10%.

913. D is correct. First of all, nitric acid is HNO_3, so choices *A* and *C* are out. Choice *D* may appear to be out, but recall that HNO_3 is a strong acid, so it dissociates completely; it is acceptable to write it as HNO_3, $H^+ + NO_3^-$, or $H_3O^+ + NO_3^-$. A closer examination of choice *B* shows that the charge is not balanced; there are three units of positive charge on the left side, and only one on the right side. Choice *D*, on the other hand, has three units of positive charge on each side.
A couple of general comments on this kind of problem: first, don't try to balance a reaction yourself when your choices are balanced reactions! It takes much less time to check whether equations are balanced correctly than to balance them yourself. In addition, sometimes the MCAT might write some piece of the reaction differently than you would, like the HNO_3 in this example. The other note is that a redox reaction is only balanced if the atoms are balanced (e.g., same number of hydrogens on each side) *and* the charge is balanced.

914. D is correct. You may either check whether each choice is balanced, as in the previous question, or you may triple the first half-reaction, double the second, and add them together (which makes the electrons cancel, as desired).

915. D is correct. It's easiest to work this problem out using electrons. Three moles of zinc, according to the first half-reaction, produce 6 moles of electrons. And 6 moles of electrons, according to the second half-reaction, can reduce 6 moles of MnO_2.

916. B is correct. It's easiest to work with electrons:

$$(3 \text{ mol } Cu^+)(1 \text{ mol } e^-/1 \text{ mol } Cu^+)(1 \text{ mol Mg}/2 \text{ mol } e^-) = 3/2 \text{ mol Mg}$$

917. C is correct. When the MCAT gives you answers in this form, do not try to work out the answer for yourself from scratch. If you do things in a different order than the question writer, you might get confused. Instead, look for what's different between the answers. Choice *A* has the 3 on top, while the others have it on the bottom. 3 is the current. With more current, should it take more or less time to reduce a fixed amount of copper? Since it should take less time, the time and current should be inversely proportional, and the 3 should be in the denominator. So choice *A* is out. The only other difference is the 2: does it belong on top, on the bottom, or nowhere? The two must be the number of electrons each copper atom is gaining. As an example, would reducing Fe^{3+} to Fe take more or less time than it would take to reduce Fe^{2+} to Fe? Since the answer is *more* time, the time must be proportional to the change in oxidation number. Thus the 2 belongs in the numerator, and *C* is the answer.

918. A is correct. Use the following calculation:

(0.108 g Ag)(1 mol Ag/108 g Al)(1 mol e^-/mol Ag)(96,485 C/mol e^-)(1 sec/2 C) = 48 seconds

919. D is correct. The Nernst equation is really about non-standard conditions, so *A* seems doubtful. Standard potential is just the potential at standard conditions. This definition does not directly involve equilibrium, so *B* seems doubtful as well. *C* is kind of silly if you think about it: standard potentials are measured under standard conditions. But *D* is a true statement: you need *two* half-reactions to make an actual reaction.

920. C is correct. Oxidation and reduction always occur together. Half reactions can never occur by themselves.

921. B is correct. Reducing agents want to be oxidized. Hg^{2+} and Au^{3+} can't be oxidized; their charges can't be increased anymore. F^- and Cl^- are already shown being oxidized, and the voltage is greater for the chloride, so choice *B* is correct. It is a fairly standard MCAT trick to write some half-reactions as oxidations rather than reductions, so watch out for it!

922. B is correct. The table shows Au^{3+} being reduced, so we have to reverse the sign.

923. B is correct. The table shows fluoride being oxidized, so we take the number directly from the table.

924. D is correct. Recall that, unlike most other metals, the money metals (Au, Pt, Ag, Hg, Cu, Ni) don't rust or tarnish because they have negative oxidation potentials. Nickel is the exception to the rule. Nickel has a slightly positive oxidation potential. Most other metals like to give up electrons; they have positive oxidation potentials. Ozone is a stronger oxidizing agent than dioxygen. Of the choices, gold has the lowest oxidation potential, so it is the most difficult to oxidize.

925. A is correct. This relationship is similar to the relationship between an acid and its conjugate base.

926. C is correct. Rusting is a form of oxidation, so the magnesium must be oxidized preferentially. Reducing agents contain the element being oxidized, so the correct answer is *C*.

927. C is correct. Potential is defined as energy per unit charge, and thus, when *I* and *II* double, *III* remains the same.

928. B is correct. Sodium is very metallic and thus very easily oxidized; fluorine is extremely nonmetallic and thus very easily reduced. Such elements are not naturally found in pure form. (An element such as gold, on the other hand, which is quite hard to oxidize, is generally found as the pure element.)

929. D is correct. Copper has a negative oxidation potential and is below hydrogen, while the other choices have a positive oxidation potential and are above hydrogen. The reduction potential of H^+ is 0 V. H^+ ions have a reduction potential of zero, so they will spontaneously reduce in the presence of a metal with a positive oxidation potential.

930. D is correct. Silver ion has the highest potential to gain an electron or take one away from something else.

931. A is correct. Lithium solid has the highest potential to lose an electron or give one away.

932. B is correct. On an activity series, a metal can reduce any metal below it. Al will loses three electrons and will not work. The most electropositive metals are generally the most easily oxidized and are thus higher in the activity series. Since magnesium is an alkaline earth, it is more electropositive than the other choices.

933. D is correct. Silver is below copper in the activity series, so silver ions can oxidize copper.

934. C is correct. Copper is more easily oxidized than silver, so, since nitric acid oxidizes silver, it will oxidize copper. The gas produced will be NO because H^+ ions will not oxidize copper.

935. C is correct. These are the properly ordered steps for balancing a redox reaction.

936. B is correct. These are the properly ordered steps for balancing a half reaction.

937. D is correct. Mn is reduced, CN is oxidized.

938. B is correct. Mn is reduced, CN is oxidized.

939. B is correct. B is the only balanced equation.

940. D is correct. D is the only balanced equation.

941. C is correct. The OH^- ions must neutralize the H^+ ions producing water. The waters cancel out on the left and appear on the right.

942. D is correct. The reduction half reaction has four H^+ ions to cancel out.

943. C is correct. Only choice C is simplified and balanced.

944. B is correct. You should try to solve this the long way by going through the steps, just for practice. Hydrogen ions and water molecules will always be on opposite sides of the reaction, so A and D are wrong. You can use plug and chug to see that the oxygens aren't balanced in C.

945. C is correct. In this problem you can narrow it down to B and C because Bi^{3+} and Na^+ must have the same number, but you have to balance the entire equation to choose between B and C.

$$14H^+(aq) + 5NaBiO_3(s) + 2Mn^{2+}(aq) \rightarrow 5Na^+(aq) + 5Bi^{3+}(aq) + 2MnO_4^-(aq) + 7H_2O(l)$$

946. A is correct. In basic solution, the OH^- ions and the water molecules are on the same side in a balanced redox reaction. But that doesn't help here. Follow the steps for balancing a redox reaction. The balanced equation is:

$$1OH^-(aq) + 5H_2O(l) + 1NO_2^-(aq) + 2Al(s) \rightarrow 1NH_3(aq) + 2Al(OH)_4^-(aq) + 0H_2O(l) + 0OH^-(aq)$$

947. D is correct. Follow the steps for balancing a redox reaction. The balanced equation is:

$$0OH^-(aq) + 0H_2O(l) + 2Cr(OH)_3(s) + 6ClO^-(aq) \rightarrow$$

$$2CrO_4^{2-}(aq) + 3Cl_2(g) + 2H_2O(l) + 2OH^-(aq)$$

948. C is correct. Galvanic cells allow the chemical energy of redox reactions to be changed to electrical energy.

949. B is correct. Electrons are negatively charged and the anode is negatively charged in a galvanic cell. The electrons are repelled by the negative charge and flow to the positively charged cathode. As they do so, they release potential energy.

950. D is correct. Negatively charged anions in the salt bridge move toward the anode to prevent a negative charge build up at the cathode as negatively charged electrons move to the cathode through the load (the resistance between the electrodes located outside the cell).

951. B is correct. Electrons flow from the anode to the cathode, so B or D must be correct. The zinc solid dissolved and Cu^{2+} formed copper solid in the first experiment, so zinc must lose electrons and copper must gain electrons.

952. D is correct. The length of the wire doesn't affect the emf of a galvanic cell.

953. A is correct. The strongest oxidizing agent "wants" to be reduced the most. The reduction potential of Cd^{2+} is –0.40 and Na^+ is –2.71. F^- and I^- cannot be reduced any further than they already are. Thus, Cd^{2+} is the strongest oxidizing agent of the choices.

954. C is correct. This time, we need an agent that "wants" to be oxidized. Na^+ and Cd^{2+} can't be oxidized any further, so they're out. The oxidation potential of F^- is –2.87 V, and that of I^- is –0.54, so I^- is the strongest reducing agent of the choices.

955. C is correct. In the table, sodium is shown being reduced and iodide is shown being oxidized. Thus, we can combine them into an oxidation-reduction reaction with a cell potential of $-0.54 + (-2.71) = -3.25$ V. But galvanic cells utilize spontaneous reactions, and spontaneous reactions have positive voltages, so the actual reaction must be the reverse reaction, with a voltage of 3.25 V.

956. C is correct. According to the explanation for the previous question, both elements must be undergoing the reverse of the half-reactions shown in the table. Reduction always takes place at the cathode, so *C* is the correct answer.

957. B is correct. Both half-reactions are written as reductions, so one must therefore be reversed. We reverse the one that will lead to a positive total when we combine the half-reactions, which in this case is the lithium half-reaction. So the total voltage is $-(-3.05\text{ V}) + (-0.76\text{ V}) = 2.29$ V. The fact that the first reaction has to be doubled to balance the reaction is irrelevant; potentials do not change when a reaction is multiplied by a constant.

958. C is correct. To understand why a salt bridge is unnecessary here, we must remember why a salt bridge is generally necessary. A salt bridge is usually required to keep one half-cell from accumulating a surplus of charge. But the battery described does not have separate half-cells, so the question becomes: How can the automobile battery function without separate half-cells? Now we can look at the choices for an explanation of this. Choice *C* fits the bill perfectly.

959. B is correct. The relationship is given by the equation shown.

960. C is correct. The equilibrium constant alone cannot indicate whether or not a reaction is spontaneous.

961. A is correct. A positive emf indicates a negative delta G. $\Delta G = -nFE$

962. A is correct. The emf is in units of volts, so it isn't even the right units for work. ΔG^o is the Gibb's energy for standard conditions only, so it doesn't cover all cells.

963. D is correct. At equilibrium, the cell can do no work, so ΔG is equal to zero. Algebraic manipulation gives $\Delta G^o = -RT \ln(K)$.

964. B is correct. The zero superscript indicates standard state which is 1 *M* for all concentrations.

965. A is correct. The zinc half reaction is reversed to make the galvanic cell, so that Zn^{2+} is a product. At standard potential the concentrations are at 1 *M* and the voltage is 1.1 V. Since we are increasing the concentration of a product, we lower the voltage from this potential.

966. C is correct. Copper is reduced in the galvanic cell, so that Cu^{2+} is a reactant. At standard potential the concentrations are at 1 *M* and the voltage is 1.1 V. Since we are increasing the concentration of a reactant, we raise the voltage from the standard potential.

967. B is correct. The Nernst equation is: $E = E^o - (0.06/n) \times \log(Q)$. $Q = [Zn^{2+}]/[Cu^{2+}] = 2/2$. The log of 1 is zero, so $E = E^o$.

968. D is correct. The voltage drop over time is represented by choice D.

969. A is correct. At standard conditions, the concentration of all reactants and products is 1 *M*, so the pH is –log 1 = 0. Thus pH 7 represents a much lower concentration of H^+ ions, and the reaction at the cathode would be considerably retarded. Thus, the voltage would decrease.

970. C is correct. Electrolytic cells are always driven by an external power source; this external source gains its power from a spontaneous reaction. The other statements are incorrect.

971. C is correct. This galvanic cell has a positive standard potential, and is thus spontaneous under standard conditions. *III* is another way of signifying this, which allows us to eliminate choice *A*. There is no particular entropy requirement for a cell to be spontaneous, however, so *II* is not necessary. That leaves us with only choice *C*.

972. A is correct. For a reaction to proceed in the forward direction, Q must be less than K. ΔG must be negative, as well, but this says nothing about $\Delta G°$; indeed, under standard conditions, this cell is nonspontaneous. Likewise, since the cell is nonspontaneous under standard conditions, *I* is false.

973. A is correct. At equilibrium, E and ΔG will be zero. E° and ΔG° will depend upon the half reactions.

974. C is correct. Positive potentials correspond to spontaneous reactions, and thus negative free energies. The "standard" in standard potential refers to the conditions. Choice *D* can be eliminated because, at equilibrium, the free energy change of any reaction is 0.

975. B is correct. The oxidant is at the cathode in a galvanic cell, so A is wrong. Changing the internal resistance won't change the emf, so C is wrong. Raising the temperature won't necessarily increase the emf, so D is wrong. B is correct based on LeChatelier's principle.

976. A is correct. This is definitional. The electrodes in an electrolytic cell have opposite signs to electrodes in a galvanic cell.

977. C is correct. You probably think of a "battery" when you think of an electrochemical cell (also known as a *galvanic cell*), so *I* must be correct. But that means either *II* or *III* must also be right (look at the answer choices). *III* is a non-spontaneous process, because gold is very stable. Thus, you're forced to choose *C*. (Or, you might know that a pH meter is based on a concentration cell, which is a special type of electrochemical cell.)

978. B is correct. This is, of course, the flip side of the previous question.

979. D is correct. The emf is an intensive property, thus volume of solution does not affect it. Temperature, concentration, and reaction determine the emf.

980. C is correct. Electrolytic cells use reactions that are forced to proceed by an external power source. They are not effective power sources themselves. *A* and *D* are standard uses of electrolytic cells. *B* is a slightly tricky choice: the electrolytic cell is the "electrochemical cell" that is being recharged; during the recharging process it is an electrolytic cell.

981. C is correct. Since electrons are flowing into the sample, the probe must be the site of reduction, i.e., the cathode. Furthermore, since the sample is receiving electrons, H^+ in the sample is being converted into H_2, raising the pH. Concentration cells are trying to equalize the concentration in the two cells, so if the pH in the sample is increasing, it must have started at a pH below 2.0.

982. D is correct. The problem appears to be at the cathode, where hydrogen is appearing rather than sodium metal. Reduction takes place at the cathode, suggesting choices *B* or *D*. Reduction of water does indeed produce hydrogen gas, since the oxidation number of H in H_2O is +1, but is 0 in H_2.

983. A is correct. The zinc electrode is not a candidate, since it is not present at the anode. Zinc ions are present, but cannot be oxidized further. That leaves only the silver electrode and the water; to determine which is oxidized, we consult the table. The oxidation potential of silver is –0.80 V; that of water is –1.23 V. Therefore, the silver is preferentially oxidized.

984. C is correct. The zinc electrode cannot be reduced any further, so *A* is eliminated. To decide among the other choices, consult the table: the reduction potential of water is –0.83 V, the reduction potential of silver ions is +0.80 V, and the reduction potential of zinc ions is –0.76 V. Since the silver ions have the greatest reduction potential, the silver ions must be reduced.

985. D is correct. You should know that silver is oxidized at the anode and then the ions reduced at the cathode. The net result is a transfer of silver from anode to cathode ("electroplating"), rather than a chemical change.

986. B is correct. Since no chemical reaction takes place, and everything is under standard conditions, no voltage is necessary. (In reality, internal resistance and the like necessitate some voltage, but the problem tells you to neglect these considerations.)

987. B is correct. Sodium is much easier to oxidize than zinc, so the anode reaction would certainly involve sodium. But, according to the table, zinc ions are easier to reduce than either sodium ions or water, and thus the zinc would be reduced at the cathode.

988. B is correct. Silver ions are no longer available at the cathode. The next highest reduction potential is for the zinc ions (but barely; it's possible a *little* bit of water would also be reduced, but we're not given this as an option.)

989. B is correct. Don't let the presence of a salt bridge distract you. The direction of electron flow is determined by an external power supply; thus, it must be an electrolytic cell.

990. C is correct. Water will be oxidized at the anode and reduced at the cathode. It takes a little effort to get everything to cancel right, but *C* is the net result. (The H^+ formed at the anode and the OH^- formed at the cathode can recombine to form water in solution.)

991. C is correct. The two half-reactions are already written in the right directions in the table, so all you have to do is add them. (Don't forget that potential is an *intensive* quantity, so it does *not* double when you double a reaction.)

992. A is correct. The standard reduction and oxidation potentials will cancel exactly.

993. A is correct. Electrons flow from the anode to the cathode in a galvanic cell. A concentration cell, which is a galvanic cell, tries to even out the concentrations. Therefore, the cation concentration in a concentration cell will always be greater at the cathode. The concentration cell will move electrons to the cathode to reduce the concentration in an attempt to even the concentrations in the half cells. This can also be seen from the Nernst equation, where Q must be less than one for a positive potential.

994. A is correct. Electrons flow from the anode to the cathode in a galvanic cell. A concentration cell, which is a galvanic cell, tries to even out the concentrations. Therefore, the cation concentration in a concentration cell will always be greater at the cathode. The concentration cell will move electrons to the cathode to reduce the concentration in an attempt to even the concentrations in the half cells.

995. B is correct. Electrons flow from the anode (producing cations at the anode) to the cathode in a concentration cell (using up cations at the cathode). The cations at the cathode are the reactants, and the cations at the anode are the products. Q is products over reactants.

996. D is correct. The zinc reduction potential is turned around to give $E^\circ = 1.56$. Silver ions are the reactants and zinc ions are the products. $Q = [Zn^{2+}] / [Ag^+]^2 = 0.1$. The log of 0.1 is negative 1. $n = 2$. So from the Nernst equation we have $E = 1.56 - (-.03) = 1.59$.

997. A is correct. Use units to solve this one. g = (197 g/mol Au x 2 c/s x 30 s)/(96,500 c/mol e^- x 3 mol e^-/mol Au)

998. C is correct. The equation is (M.W.)(i)(t)/(F)(n) = (m) as derived in the previous problem. The emf doesn't enter into the equation because the problem states that the current is the same. Gold is three times heavier but requires 3 moles of electrons per mole of gold, where copper only requires 2 moles of electrons per mole of gold. Thus copper will acquire 1.5 as many moles in the same amount of time, but gold still weighs three times as much. When the 1 gram of gold is plated, only ½ gram of copper will have been plated.

999. C is correct. Since zinc is oxidized more easily than iron, when the surface coat is broken, zinc will act as the anode and corrode instead of the iron.

1000. A is correct. Oxidation occurs where something loses electrons. The iron dissolves into solution at electrode Y. Corrosion starts with solid iron. You can even see that the metal is eaten away at electrode Y.

1001. D is correct. Oxygen is reduced at electrode X.